T-Labs Series in Telecommunication Services

Series editors

Sebastian Möller, Berlin, Germany
Axel Küpper, Berlin, Germany
Alexander Raake, Berlin, Germany

More information about this series at http://www.springer.com/series/10013

Shahin Tajik

On the Physical Security of Physically Unclonable Functions

 Springer

Shahin Tajik
Telekom Innovation Laboratories
Technical University of Berlin
Berlin
Germany

ISSN 2192-2810 ISSN 2192-2829 (electronic)
T-Labs Series in Telecommunication Services
ISBN 978-3-030-09333-4 ISBN 978-3-319-75820-6 (eBook)
https://doi.org/10.1007/978-3-319-75820-6

Printed on acid-free paper

This Springer imprint is published by the registered company Springer International Publishing AG
part of Springer Nature
The registered company address is: Gewerbestrasse 11, 6330 Cham, Switzerland

Dedicated to my family

List of Publications

The primary results of this work have been presented in the following publications:

- **Tajik, S.**, Nedospasov, D., Helfmeier, C., Seifert, J.-P., Boit, C.: *Emission Analysis of Hardware Implementations*. In Proceedings of 17th Euromicro Conference on Digital System Design (DSD), IEEE, 2014, Verona, Italy
- **Tajik, S.**, Dietz, E., Frohmann, S., Seifert, J.-P., Nedospasov, D., Helfmeier, C., Boit, C., Dittrich, H.: *Physical Characterization of Arbiter PUFs*. In Proceedings of 16th International Workshop on Cryptographic Hardware and Embedded Systems—CHES 2014, Busan, South Korea
- **Tajik, S.**, Lohrke, H., Ganji, F., Seifert, J.-P., Boit, C.: *Laser Fault Attack on Physically Unclonable Functions*. In Proceedings of Workshop on Fault Diagnosis and Tolerance in Cryptography (FDTC), IEEE, 2015, St. Malo, France
- Lohrke, H., **Tajik, S.**, Boit, C., Seifert, J.-P.: *No Place to Hide: Contactless Probing of Secret Data on FPGAs*. In Proceedings of 18th International Conference on Cryptographic Hardware and Embedded Systems—CHES 2016, Santa Barbara, USA
- **Tajik, S.**, Dietz, E., Frohmann, S., Dittrich, H., Nedospasov, D., Helfmeier, C., Seifert, J.-P., Boit, C., Hübers, H.W.: *Photonic Side-channel Analysis of Arbiter PUFs*. Journal of Cryptology, Springer-Verlag, 2017
- **Tajik, S.**, Fietkau, J., Lohrke, H., Seifert, J.-P., Boit, C.: *PUFMon: Security Monitoring of FPGAs using Physically Unclonable Functions*. accepted for 23rd IEEE International Symposium on On-Line Testing and Robust System Design (IOLTS), IEEE, 2017, Thessaloniki, Greece

Additionally, Shahin Tajik has authored the following publications:

- Helfmeier, C., Nedospasov, D., **Tajik, S.**, Boit, C., Seifert, J.-P.: *Physical Vulnerabilities of Physically Unclonable Functions*. In Proceedings of Design, Automation, and Test in Europe Conference and Exhibition—DATE 2014, Dresden, Germany

- Ganji, F., **Tajik, S.**, Seifert, J.-P.: *Why Attackers Win: On the Learnability of XOR Arbiter PUFs*. In Proceedings of 8th International Conference on Trust and Trustworthy Computing—TRUST 2015, Heraklion, Greece
- Ganji, F., **Tajik, S.**, Seifert, J.-P.: *Let me prove it to you: RO PUFs are provably learnable*. In Proceedings of 18th Annual International Conference on Information Security and Cryptology (ICISC), 2015, Busan, South Korea
- Ganji, F., Krämer, J., Seifert, J.-P., **Tajik, S.**: *Lattice Basis Reduction Attack against Physically Unclonable Functions*. In Proceedings of 22nd ACM Conference on Computer and Communications Security—CCS 2015, Denver, USA
- Ganji, F., **Tajik, S.**, Seifert, J.-P.: *PAC learning of arbiter PUFs*. Journal of Cryptographic Engineering, Springer-Verlag, 2016
- Boit, C., **Tajik, S.**, Scholz, P., Amini, E., Beyreuther, A., Lohrke, H., Seifert, J.-P.: *From IC Debug to Hardware Security Risk: The Power of Backside Access and Optical Interaction*. In Proceedings of 23rd International Symposium on the Failure Analysis of Integrated Circuits—IPFA 2016, Marina Bay Sands, Singapore
- Ganji, F., **Tajik, S.**, Fäßler, F., Seifert, J.-P.: *Strong Machine Learning Attack Against PUFs with No Mathematical Model*. In Proceedings of 18th International Conference on Cryptographic Hardware and Embedded Systems—CHES 2016, Santa Barbara, USA
- Lohrke, H., **Tajik, S.**, Scholz, P., Boit, C., Seifert, J.-P.: *Automated Detection of Fault Sensitive Locations for Reconfiguration Attacks on Programmable Logic*. In Proceedings of 42nd International Symposium for Testing and Failure Analysis—ISTFA 2016, Fort Worth, USA
- Ganji, F., **Tajik, S.**, Fäßler, F., Seifert, J.-P.: *Having No Mathematical Model May Not Secure PUFs*. Journal of Cryptographic Engineering, Springer-Verlag, 2017

Contents

Abbreviations

AES	Advanced Encryption Standard
AI	Artificial Intelligence
APD	Avalanche Photodiode
ASIC	Application-Specific Integrated Circuit
BBRAM	Battery-Backed Random Access Memory
BGA	Ball Grid Array
BR	Bistable Ring
BS	Beam Splitter
CB	Control Box
CCD	Charged-Coupled Device
CED	Concurrent Error Detection
CMOS	Complementary Metal-Oxide Semiconductor
CPLD	Complex Programmable Logic Device
C-RAN	Centralized-Radio Access Network
CRP	Challenge Response Pair
DPA	Differential Power Analysis
DUT	Device under Test
DWC	Duplication with Comparison
EM	Electromagnetic
FFT	Fast Fourier Transform
FIB	Focused Ion Beam
FPGA	Field Programmable Gate Array
FSBL	First Stage Boot Loader
IC	Integrated Circuit
IDE	Integrated Development Environment
IoT	Internet of Things
IP	Intellectual Property
JTAG	Joint Test Action Group
LAB	Logic Array Block
LE	Logic Element

LFI	Laser Fault Injection
LSB	Least Significant Bit
LSM	Laser Scanning Microscope
LUT	Lookup Table
LVI	Laser Voltage Imaging
LVP	Laser Voltage Probing
ML	Machine Learning
MOS	Metal-Oxide Semiconductor
MSB	Most Significant Bit
NA	Numerical Aperture
NIR	Near-Infrared
nm	nanometer
NVM	Nonvolatile Memory
PAC	Probably Approximately Correct
PCB	Printed Circuit Board
PEM	Photonic Emission Analysis
PLD	Programmable Logic Device
PLS	Photoelectric Laser Stimulation
PoC	Proof of Concept
PRNG	Pseudo-Random Number Generator
ps	picosecond
PUF	Physical(ly) Unclonable Function
rms	root mean square
RO	Ring-Oscillator
RoT	Root of Trust
SCA	Side-Channel Analysis
SDN	Software Defined Network
SDR	Software Defined Radio
SEE	Single Event Effects
SEM	Scanning Electron Microscope
SFD	Summation of the Frequency Differences
SoC	System on Chip
SPAD	Single-Photon Avalanche Diode
SRAM	Static Random Access Memory
TDC	Time-to-Digital Converter
TLS	Thermal Laser Stimulation
TMR	Triple Modular Redundancy
TQFP	Thin Quad Flat Package
TRNG	True Random Number Generator
XOR	Exclusive Or

List of Figures

List of Tables

Abstract

Reconfigurable hardware is the primary component of electronic embedded devices employed in several applications ranging from wireless communication to cloud computing. Due to their significant role, these modern platforms are targets of intellectual property (IP) piracy and tampering. Cloning of a design or manipulation of its content is carried out by conducting physical attacks (e.g., side-channel analysis and fault attacks) against these devices. Although different countermeasures against physical attacks have been integrated into the modern reconfigurable hardware, a proper protection mechanism on these platforms against semi-invasive attacks conducted from the chip backside is still missing. The main and foremost reason that the chip backside protection is ignored by the vendors is the misconception that semi-invasive attacks cannot be scaled to the very latest nanoscale technologies without further effort and cost. Moreover, it is assumed that integrating novel hardware intrinsic-based solutions for key storage, such as Physically Unclonable Functions (PUFs), make the conventional semi-invasive memory readout techniques virtually impossible.

In this work, we investigate the susceptibility of Intrinsic PUF implementations on reconfigurable hardware to optical semi-invasive attacks from the chip backside. We conduct different classes of optical attacks, particularly photonic emission analysis, laser fault injection, and optical contactless probing. By applying these techniques, we demonstrate that the secrets generated by a PUF can be predicted, manipulated, or directly probed without affecting the behavior of the PUF. We further discuss the cost and feasibility of launching such attacks against the very latest hardware technologies in a real scenario. We discuss why PUFs are not tamper-evident in their current configuration, and therefore, PUFs alone cannot raise the security level of key storage, as one would expect in the first place. Moreover, we review the potential and already realized countermeasures, which can remedy the security-related shortcomings of the PUFs and make them resistant to optical side-channel and optical fault attacks. Finally, by making a few modifications in the functionality of an existing PUF architecture, we present a prototype of a tamper-evident sensor for detection of optical contactless probing attempts.

Chapter 1
Introduction

1.1 Motivation and Background

Modern electronic embedded devices have become indispensable parts of our daily lives. End user devices, such as smartphones, smartwatches, and smart home appliances, gather data in an unprecedented way and make ubiquitous computing feasible. Moreover, industrial internet of things (IoT) consisting of robots, industrial controllers, and smart grids is the essential part of our modern infrastructure. More data accumulations at the edge are meaningless without deep learning and analysis of them in the cloud. Needless to say that devices at the edge communicate with data centers over a low-latency and high-speed telecommunication infrastructure comprised of routers and switches. While benefits of ubiquitous computing in our lives are indisputable, there are several concerns regarding the security of deployed electronic embedded devices in all parts of this huge network. A wide variety of utilized embedded devices in consumer, industrial and military applications are targets of reverse-engineering and Intellectual Property (IP) piracy. The primary motivation behind reverse-engineering is to get access to the stored secrets and utilized IPs on the integrated circuits (ICs) to counterfeit and overbuild the target products [86, 91]. Consequently, a great deal of attention has to be paid to protect the employed secrets and IPs on the embedded devices.

1.1.1 Reconfigurable Hardware

Embedded systems can be developed either by software or hardware implementations. While in the former case the desired functionality is realized by running software on a microprocessor, in the latter case it is realized by application-specific integrated circuits (ASICs). Although developing software for a microprocessor is faster, updatable and less expensive in comparison to designing ASICs, the performance and the power efficiency of microprocessors are inferior to ASICs. A third

© Springer International Publishing AG, part of Springer Nature 2019
S. Tajik, *On the Physical Security of Physically Unclonable Functions*, T-Labs
Series in Telecommunication Services, https://doi.org/10.1007/978-3-319-75820-6_1

alternative for developing embedded systems is reconfigurable hardware, which combines the advantages of both software and hardware implementations. Complex Programmable Logic Devices (CPLDs) and Field Programmable Gate Arrays (FPGAs) are the most popular instances of reconfigurable hardware. These platforms can realize circuits from hundreds to millions of Boolean gates. CPLDs contain fewer configurable logic resources than FPGAs, and therefore, are preferred for realizing of lightweight applications. FPGAs, on the other hand, include conventionally more resources and are employed for applications, which require substantial processing powers. Furthermore, to get even more powerful platforms, FPGA vendors have made programmable System on Chips (SoCs), in which microprocessors are integrated into the FPGA fabrics.

Digital signal processing, software-defined radios (SDRs), and cryptography are a few examples of standard FPGA applications. Programmable SoCs are now integrated into the switches of software-defined networks (SDNs) to keep pace with changes in standards and protocols. Furthermore, they are considered as the core of centralized radio access network (C-RAN) in the 5G cellular networks. Embedded vision is another application of these platforms, which is helpful in autonomous cars, medical imaging, and video surveillance. More recently, internet giants start to integrate FPGAs and programmable SoCs into their cloud computing platforms to adapt their designs continuously to new artificial intelligence (AI) algorithms [1, 56].

Desired functionalities are configured and reconfigured on CPLDs, and FPGAs by a binary data called *bitstream*. In contrast to CPLDs, most FPGAs do not contain any non-volatile memory (NVM) inside their packages. Due to the lack of an internal NVM inside most of these devices, they cannot store the bitstream internally. Hence, a bitstream has to be kept in an external NVM and sent to the FPGA in an untrusted field upon each power-on. The transmission of a bitstream in an adversarial environment can expose the design if no protection is provided. Bitstream encryption is a conventional countermeasure provided by the FPGA vendors to keep the design confidential [88]. In this case, available Battery Backed RAMs (BBRAMs) and eFuses on the FPGAs can be used to store the secret key for the decryption of the bitstream.

1.1.2 Physical Attacks

To reverse-engineer a running application on a reconfigurable hardware, the attacker needs to reconstruct the implemented design. However, in a real attack scenario, the attacker might have access to neither the hardware description language (HDL) code nor the gate level netlist of application. Furthermore, the bitstream is not available to the attacker, since either the bitstream is transferred encrypted between NVM and FPGA or it is kept in the internal NVM of the CPLDs and flash-based FPGAs. Thus, to get access to the design, the attacker needs to launch physical attacks.

Side-channel analysis (SCA) [52], fault injection attacks [9], and microprobing of the secret [33] are a few examples of the physical attacks, which can be conducted against embedded devices. SCA exploits the leakage of information during a cryptographic operation or key generation to extract the secret key. Power analysis and Electromagnetic (EM) analysis are examples of SCA methods, in which the attacker measure the electrical current and EM radiations from the chip, respectively. In the case of fault injection attacks, the attacker attempts to observe a faulty cryptographic operation or key generation by feeding the chip with faulty data or forcing it to operate in a non-valid condition. For instance, by varying the supplied voltage (i.e., voltage glitching), altering the frequency of the clock signal (i.e., clock glitching), or flipping bits in the memory with a laser beam (i.e., laser fault injection) the attacker can cause an erroneous operation of the target device.

A set of SCA and fault injection attacks can be carried out in a non-invasive way, i.e.; it is not required to remove the package of the IC under attack and only access to the pins of the device is required, which makes these attacks inexpensive. On the contrary, the first step to conducting an invasive attack is to remove the plastic or metallic package of the IC. Microprobing attacks enable the attacker to make a direct physical contact with the transistors and wires to extract the secrets. In this case, additional to the depackaging the barriers for a physical contact, namely the metal layers on the frontside of the IC or part of the silicon substrate on the IC backside, must be removed. Hence, these attacks are considered as fully-invasive techniques. However, to perform optical attacks, such as photonic emission analysis (PEM) or laser fault injection (LFI), no physical contact with the transistors is necessary. Although in principle semi-invasive attacks can be carried out from both frontside and backside of the IC, the existing multiple interconnected layers on the frontside of the modern ICs obstruct the optical paths from transistors to the surface of the device. This fact makes the analysis of the target IC from its backside more attractive to the attacker. In this fashion, if the proper photon wavelengths are deployed, thinning or polishing of the silicon substrate is not necessary anymore. Hence, these techniques are considered as semi-invasive.

1.1.3 Physically Unclonable Function

Physically Unclonable Functions (PUFs) [28, 66] are introduced to mitigate the vulnerabilities of common key storage and key generation technologies to physical attacks on integrated circuits (ICs). On the one hand, PUFs exploit the existing manufacturing process variations on the chips to create virtually unique fingerprints, which can be utilized for the device authentication, and consequently, preventing the counterfeiting. On the other hand, these variations can be used as an entropy source to generate keys for cryptographic primitives. In general, PUFs can be thought of as mappings, which produce a *response* for a given *challenge*. Using PUFs eliminate the need for an NVM, since no secret key is required to be permanently stored on the chip. In other words, a secret key or signature is generated by feeding the PUF with a set of challenges and activating it.

Among different classes of the PUFs, *Intrinsic* PUFs [48] can be effectively implemented on reconfigurable hardware without any extra manufacturing costs using the existing resources on these platforms. Bistable PUFs and delay-bases PUFs are the two main categories of Intrinsic PUFs. While the former group exploits the metastability of the bistable circuits (e.g., SRAM cells) as a fingerprint or an entropy source, the latter group utilizes the propagation delay differences of the symmetrically designed circuits on the chip to generate randomness and unique signatures.

An ideal PUF must have several features [48], including *unclonability*, *unpredictability*, and *tamper-evidence* (see Sect. 2.1 for more details). While unclonability deals with the assumption that the behavior of a PUF is neither physically nor mathematically clonable, unpredictability deals with the inability of the attacker to predict the response of the PUF by observing a set of challenge-response-pairs (CRPs). Moreover, tamper-evidence means that if the attacker launches a semi- or fully-invasive attack against a PUF, the challenge-response behavior of the PUF is altered with a high probability leading to the destruction of PUF and loss of secret key or fingerprint.

PUFs can be employed on reconfigurable hardware to generate a key for decryption of an encrypted bitstream and bind it to a unique hardware. Besides, it can be used as the fingerprint of the reconfigurable hardware for authenticate purposes. The designers can deploy PUFs for their security applications as well, by including a PUF configuration in the bitstream.

1.2 Problem Statement

It has been demonstrated that the bitstream encryption of different series of FPGAs are vulnerable to the SCA, and therefore, the secret key can be successfully recovered [38, 60, 61]. If no proper integrity checking mechanism is implemented for the encrypted bitstreams, bitstreams are vulnerable to fault injections as well, as the adversary can flip any arbitrary bits of the bitstream to inject a fault into the computation of the application [80]. Moreover, BBRAMs and eFuses of SRAM-based FPGAs, as well as flash memory of CPLDs and flash-based FPGAs could be vulnerable to semi- and fully-invasive attacks [88]. Side-channel resistant decryptors, key rolling techniques, and asymmetric authentication schemes have been implemented on the most recent generations of FPGAs to assure the confidentiality and integrity of the bitstream against side-channel and fault injection attacks [47, 67]. Additionally, proprietary soft security monitoring IPs provided by FPGA vendors utilize the dedicated sensors inside the FPGAs to monitor the integrity of the device during runtime [58, 92]. However, a proper physical protection on the backside of these modern platforms is still missing to prevent semi- and fully-invasive attacks.

There are good reasons for reconfigurable hardware vendors to be less concerned about the security of the IC backside. First, the latest generations of the SRAM-based FPGAs are manufactured with 16 and 14 nm process technologies [19, 32]. It is believed that the conventional invasive attacks from the IC backside have limited resolution and cannot be scaled efficiently along with the trend of shrinking transistor technologies. However, in parallel to shrinking size of transistors high-resolution techniques have been developed in the failure analysis to debug the nanoscale technologies. This novel techniques can be deployed to mount an attack against the modern ICs.

Second, it is believed that PUFs raises the security level of the key storage on the reconfigurable hardware against invasive attacks [46, 57, 67] since no secret key is permanently stored on the chip to be read out by the attacker. Furthermore, it is assumed that PUFs are tamper-evident and any invasive attempt to characterize the PUF or probe the PUF responses destroys the PUF, which leads to the loss of the secret. The latter assumption is proved to be invalid since a set of attacks in the literature has been reported, which could break the security of a set of PUF architecture by semi-invasive and fully-invasive attacks [34, 54, 64].

Finally, it is assumed that semi-invasive attacks are much more expensive than other classes of attacks regarding equipment's cost and the required time for reverse-engineering Hence, vendors do not consider these attacks as a real threat to their products. While it is true that these attacks are more expensive than other conventional attacks, their cost is generally overestimated. In the literature different semi-invasive experimental setups have been demonstrated, whose costs are affordable for small-scale organizations. It is worth to mention the required setup can be rented from failure analysis labs or universities with an operator with a few hundred dollars per hour. Accordingly, these types of equipment are not necessarily to be possessed by the attacker.

1.3 Scientific Contribution

The aim of this work is to demonstrate that only replacing the NVMs (i.e., flash memories, eFuses, and BBRAMs) with PUFs does not raise the security level of the reconfigurable hardware being vulnerable to semi-invasive attacks from the IC backside. In this work, we evaluate the security of different PUF instances in two separate scenarios: (i) a PUF instance is designed by the user in HDL as a part of a larger application and configured in the reconfigurable hardware via bitstream, (ii) a PUF is deployed either by the vendor or the user inside the FPGA to store part of a secret key, which is utilized to decrypt the encrypted bitstream. In the former scenario, we assume that the adversary has access to an open interface, where she can feed arbitrary challenges to the PUF and read out the responses of it non-invasively, In the latter scenario, no electrical and non-invasive access to the challenges and responses of the PUF is available to the attacker.

For the first scenario, we present two attacks against delay-based PUFs applying PEM and LFI techniques. First, we deploy the PEM technique to characterize and predict the behavior of a delay-based PUF. Next, we demonstrate how the LFI technique can be employed to launch a modeling attack against a delay-based PUF, which is resistant to modeling attacks. Besides, we show how an attacker can adversely affect the entropy of the responses of a PUF with the help of LFI. For the second scenario, we present an attack based on optical contactless probing, which enables the attacker to probe the responses of an arbitrarily chosen PUF. We evaluate the feasibility of all attacks by conducting them against Proof-of-Concept (PoC) PUF implementations in different reconfigurable hardware.

Based on the achieved results, we review possible protection schemes against the mounted attacks. Fortunately, in response to our presented PEM and LFI attacks effective countermeasures have been proposed and realized in the literature to protect the PUF. Moreover, we discuss why PUFs, in their current shapes, cannot be deployed as anti-tamper sensors against semi-invasive attacks. Hence, we propose a small modification to one of the existing Intrinsic PUF architectures to make it tamper-evident against optical contactless probing.

1.4   Thesis Structure

This thesis is organized as follows: Chapter 2 presents background information on the PUFs, reconfigurable hardware, and semi-invasive attacks. Moreover, the related work is reviewed. In Chap. 3, the utilized experimental setups for different experiments are presented. Chapter 4 introduces the PEM attack against the delay-based PUFs. Chapter 5 presents the LFI attack against the delay-based PUFs. In Chap. 6 an attack on PUFs based on optical contactless probing is introduced. Furthermore, the prototype of a PUF-based sensor for detection of optical probing attacks is presented. Finally, in Chap. 7 we conclude the thesis and provide insights for future work.

Chapter 2
Background

In this chapter, we first review the definition of an ideal Physically Unclonable Function (PUF) and explore the functionality of two popular delay-based PUFs, namely Arbiter PUF and ring-oscillator (RO) PUF. Second, we study the internal architecture of reconfigurable hardware and discuss the security issues of FPGAs during configuration. Moreover, we survey the PUF implementations on these platforms. Afterward, we review the optical semi-invasive techniques, which are used in this work to attack PUF implementations on reconfigurable hardware. Finally, we review the related work in the literature. All provided background information in this chapter is previously presented in [45, 81–85].

2.1 Physically Unclonable Functions

Here we briefly describe the essential characteristics of an *ideal* PUF, and for a formal foundation and the formalization of the security of PUFs, we refer the reader to [8].

Definition 1 Let $\mathcal{C} = \{0, 1\}^n$ and $\mathcal{Y} = \{0, 1\}$ be the set of challenges and the set of responses, respectively. A PUF can be represented by the function $f_{\mathrm{PUF}} : \mathcal{C} \rightarrow \mathcal{Y}$ where $f_{\mathrm{PUF}}(c) = y$, cf. [50]. Note that f_{PUF} is not a one-to-one mathematical function. Ideally, f_{PUF} aims to provide the following security-related properties.

1. *Evaluable*: f_{PUF} can be evaluated in polynomial time.
2. *Unique*: for a given PUF instance, the mapping f_{PUF} is instance-specific.
3. *Reproducible*: applying same challenges to f_{PUF} results in "close" responses with respect to a chosen distance metric.
4. *Unclonable*: for a given PUF f_{PUF} it is (almost) impossible to construct another mapping (i.e., physical entity) g_{PUF} so that "$g_{\mathrm{PUF}} \approx f_{\mathrm{PUF}}$".

© Springer International Publishing AG, part of Springer Nature 2019
S. Tajik, *On the Physical Security of Physically Unclonable Functions*, T-Labs
Series in Telecommunication Services, https://doi.org/10.1007/978-3-319-75820-6_2

5. *Unpredictable*: for a given set $U = \{(c_i, y_i) \mid y_i = f_{\text{PUF}}(c_i)\}$, it is (almost) impossible to predict a response $y_r = f_{\text{PUF}}(c_r)$, where c_r is a random challenge and $(c_r, y_r) \notin U$.

6. *One-way*: for a given random PUF instance $y = f_{\text{PUF}}(c)$, where c is drawn from a uniform distribution on $\{0, 1\}^n$, we have

$$\Pr[\mathcal{A}(f_{\text{PUF}}(c)) = c] < 1/p(n),$$

where $p(\cdot)$ is any positive polynomial. In other words, it is *hard* to find c, if the respective response of a random instance of the PUF family is known, and the adversary can evaluate the PUF a polynomial number of times [48].

7. *Tamper evident*: physical altering or modifications of the physical entity embedding f_{PUF} transforms it to f'_{PUF} such that with high probability $f'_{\text{PUF}} \neq f_{\text{PUF}}$.

The current types of PUFs can only partially fulfill the requirements mentioned above. Many different PUF architectures have been proposed in the literature. *Intrinsic* PUFs [48] are one of the primary and popular classes of the PUFs since they can be effectively implemented on the embedded devices without any additional manufacturing processes. Intrinsic PUFs can generally be classified into two distinct classes: bistable PUFs and delay-based PUFs [48]. The former is based on bistable circuits such as SRAM memory cells, while the latter relies on intrinsic differences in propagation delays of symmetrically designed wires and transistors on the chip. Arbiter PUF and RO PUF are two widely deployed delay-based PUFs in the reconfigurable hardware.

2.1.1 Arbiter PUF

Arbiter PUF families exploit the slight delay differences of two symmetrically designed paths on the chip to generate binary responses [41]. A single Arbiter PUF consists of multiple serially connected stages and an Arbiter at the end of the chain, see Fig. 2.1. Each stage in this architecture contains two signal outputs, two signal inputs, and a single challenge input. The inputs of the first stage of the Arbiter PUF are connected to a common enable signal. By giving an electrical pulse as an

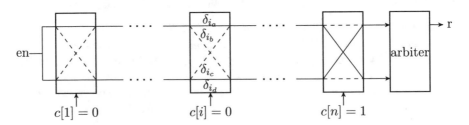

Fig. 2.1 Schematic of an Arbiter PUF

enable signal, two signals propagate on two similar paths to the end of the chain. The signal propagates through the crossed paths inside a stage if the challenge input is set to 1. Otherwise, the direct paths are utilized. Although the nominal delays of direct paths and crossed paths are equal (i.e., $\delta_{i_a} = \delta_{i_d}$ and $\delta_{i_b} = \delta_{i_c}$) the propagation delay on one of the paths can be longer or shorter due to imperfections on the chip. Hence, different challenges result in different propagation delays at the outputs of the last stage. Finally, regarding the arrival time of the signals on the outputs of the final stage, the Arbiter generates a binary response. The primary physical security assumption of an Arbiter PUF is that an attacker cannot measure the internal delays within the stages of the Arbiter PUF without destroying the PUF itself, i.e., changing its challenge-response behavior. In this case, the attacker can only try challenges from an exponential space and observe the respective responses.

Due to its relatively large challenge space, Arbiter PUFs are considered promising candidates for authentication protocols [20]. However, it is known that Arbiter PUFs are vulnerable to machine learning (ML) attacks [41]. An attacker can model the internal delays of an Arbiter PUF by applying an ML algorithm to a set of CRPs. As a countermeasure, the XOR Arbiter PUF has been introduced to impair the effectiveness of ML attacks [79]. An XOR Arbiter PUF has k parallel Arbiter chains, each with n stages and an Arbiter, see Fig. 2.2. The joint binary response is generated by XOR-ing the responses of all individual Arbiter chains. In this case, only the joint

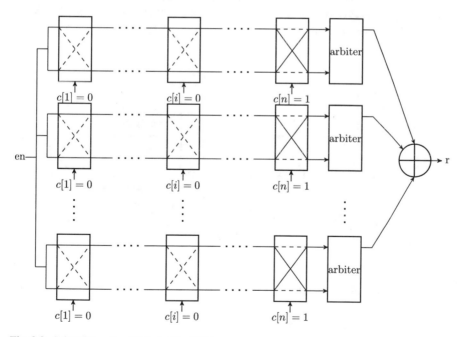

Fig. 2.2 Schematic of an XOR Arbiter PUF

response is available, and the responses of each single Arbiter PUF are hidden from the attacker. XOR Arbiter PUFs cannot be learned in a polynomial time if the number of Arbiter PUFs in this architecture are larger than a threshold [25].

2.1.2   Ring-Oscillator PUF

RO PUFs [79] are another delay-based PUFs exploiting the intrinsic timing differences on the chip. A ring-oscillator circuit consists of an odd number of inverters gates and optionally an AND gate to activate or deactivate the circuit. An RO PUF consists of n independent ring-oscillators with the same number of gates, where all ring-oscillators are connected to an n-to-2 multiplexer, see Fig. 2.3. Although all ring-oscillators have the same lengths, their oscillation frequencies are slightly different due to the manufacturing process variations on the chip. By applying challenges to the multiplexer, a ring-oscillator pair is selected, and their outputs will be connected directly to the clock inputs of 2 binary counters. The counters count the number of the rising edges of the signals at the outputs of the ring-oscillators. Because of a difference in frequencies, counters deliver different values after a predefined period. Finally, a comparator compares the states of the counters and generate a binary response. The central physical security assumption of an RO PUF is that an attacker cannot precisely measure the oscillation frequencies of ring-oscillators and predict the outputs of the PUF.

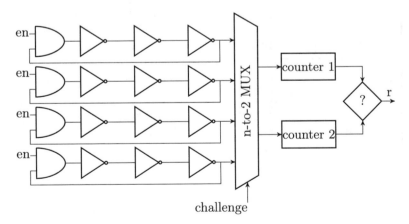

Fig. 2.3 Schematic of an RO PUF

2.1.3 Are PUFs Tamper-Evident?

PUFs are believed to be tamper-evident against invasive attacks. Being tamper-evident against fully-invasive attacks have been experimentally verified for optical and coating PUFs [68, 89]. However, they cannot be integrated into most platforms without additional manufacturing steps, and therefore, they are not considered as Intrinsic PUFs [48]. Unfortunately, for Intrinsic PUFs, limited information on tamper-evidence is available in the literature. Fortunately, results on effects of semi-invasive analysis on circuits similar to delay-based PUFs can be found in the literature related to the failure analysis. It has been shown that mechanical stress from depackaging and substrate thinning have negligible effects on the absolute and relative frequencies of ring-oscillators [14]. In another experiment, it has been reported that removing most of the bulk silicon, down to the bottom of the n-wells, does not alter the delays of the inverter chains [74]. Moreover, different successful semi-invasive attacks have been reported on silicon intrinsic PUF instances without affecting the challenge-response behavior of the PUFs [34, 54, 64]. On the other hand, PUF developers do their best to mitigate the noisy responses of the PUF by different error correction techniques [37, 49]. Hence, if the physical tampering changes a few CRPs, they will be corrected by error correction methods. Consequently, semi-invasive attacks do not destroy the Intrinsic PUFs.

2.2 Reconfigurable Hardware

Programmable Lookup Tables (LUTs), registers, and routing switches are the basic components of programmable logic devices. Several LUTs, each with multiple inputs and one output, are responsible for the configuration of combinatorial logic functions. A 4-input LUT, which contains 16 SRAM cells can be seen in Fig. 2.4. The inputs of the LUT can be regarded as the addresses of SRAM cells, which access the stored values by multiple multiplexers. To realize sequential logic functions dedicated programmable registers can be utilized. Furthermore, routing of different signals between LUTs and registers is configured by programmable switches on the chip. CPLDs and FPGAs are popular instances of reconfigurable hardware. The architectures of modern CPLDs and FPGAs are virtually similar. However, the main architectural differences between these two devices are related to their logical sizes, routing complexities and dedicated peripherals. Generally, CPLDs contain less logical resources than FPGAs, and hence, are preferred for less complex computing applications.

Reconfigurable hardware is programmed and configured by a bitstream, which is generated by an application designer. While CPLDs and flash-based FPGAs have internal NVM to store configuration data in the same package, SRAM-based FPGAs do not contain any NVM inside the package, and therefore, are not capable of storing the bitstream [88]. Therefore, the bitstream has to be stored in an external NVM

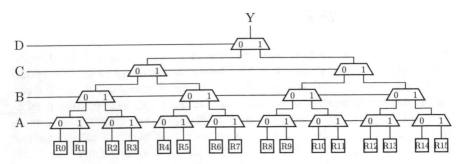

Fig. 2.4 Schematic of a 4-input lookup table

and loaded into the SRAM-based FPGAs upon each power-on in an adversarial environment. Even if the firmware of CPLDs and flash-based FPGAs require updates, their upgraded bitstream has to be transmitted from an external NVM to the device in an untrusted field. Transferring bitstreams in plaintext can divulge the designs and IPs to an adversary. As a result, bitstreams have to be kept confidential.

2.2.1   Security During Configuration

Bitstream encryption is a conventional solution to prevent the IP piracy during FPGA configuration. To enable the bitstream encryption a secret *red key* (i.e., an unencrypted key) is transferred to the FPGA in a safe environment, see Fig. 2.5a. This key will be stored either in the BBRAM or eFuses inside the FPGA. At the same time, the application design is encrypted in the integrated development environment (IDE) software by the red key and stored in an external NVM. Each time the FPGA is powered up in the untrusted field, the encrypted bitstream is transmitted to the chip, and it will be decoded by a decryption core and the stored red key inside the chip. Although this technique raises the security of the bitstream transmission against interception, it has been demonstrated that the decryption cores on different FPGAs are vulnerable to EM and differential power analysis (DPA) [38, 60, 61]. Moreover, the key storage technologies on FPGAs such as eFuses are susceptible to semi-invasive attacks and can be read out with a scanning electron microscope (SEM) [88].

Utilizing updatable protected soft decryption cores, asymmetric authentication, and key rolling can defeat non-invasive side-channel attacks, such as DPA and EM analysis [67]. Moreover, PUFs [28, 66] can remedy the shortcomings of insecure storage in modern FPGAs [88]. PUFs can be used for secure key generation and key obfuscation in an untrusted environment, where the adversary has access to the device and is able to mount a physical attack. Additionally, PUFs can be deployed as unique identifiers to prevent cloning and spoofing [29, 30, 47, 77, 96]. In the latter case, the used IPs in the application design can be coded to operate only on a specific device.

PUF and DPA-resistant decryptors can be realized either by dedicated logic inside the FPGA in the form of ASIC (i.e., hard cores) or by configuring the FPGA logic cells (i.e., soft cores). Although the principle of using PUFs for key obfuscation and DPA-resistant decryptors to defeat SCA are similar among different FPGA vendors, the implementation details differ. In this work, we explain the red key wrapping technique using soft PUFs and soft decryptors, which is deployed by Xilinx SoCs [67]. In the trusted field a boot loader containing the red key and a soft PUF IP is transferred into the volatile configuration SRAM of the FPGA. After the boot loader is loaded, the PUF is configured on the programmable logic of the device, and its responses are used in conjunction with the red key to generate the *black key* [67], see Fig. 2.5b. The black key generated in this way can only be converted back to the red key with the correct, chip-specific, internal-only PUF response (i.e., PUF key). Hence, the black key can then be stored safely in an insecure NVM, and the red key will only exist as volatile, internal-only data.

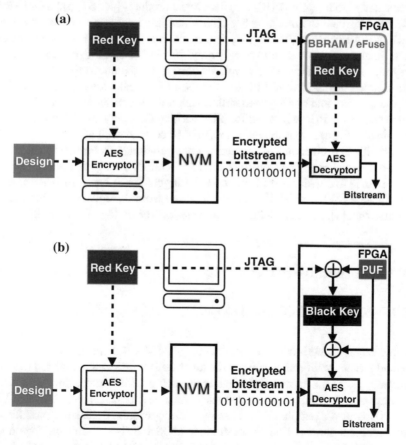

Fig. 2.5 **a** Bitstream encryption and decryption using a red key [88]. **b** Bitstream encryption and decryption using a black key, PUF key and red key [67]

In the untrusted field, an encrypted first stage boot loader containing the black key, the same soft PUF IP, and a DPA-resistant decryption IP core is loaded into the device. The chip-specific PUF response is then used to unwrap the black key and generate the red key on the fly. In the second phase, the encrypted configuration bitstream is transferred to the device and will be decrypted by the red key inside the FPGA. In this case, the decryption IP core can be updated against future side-channel analysis threats. Furthermore, the soft PUF in conjunction with the black key provides volatile, internal-only and updatable key storage, and therefore, the red key is in memory only during the configuration of the device.

2.2.2 PUF Implementations on FPGAs

Current FPGA market leaders, namely Xilinx, Intel/Altera and Microsemi, have already started to integrate PUFs into their latest products [46, 57, 67]. Hard SRAM PUFs from Intrinsic-ID Inc. have been integrated into the Microsemi SmartFusion2, IGLOO2 and PolarFire FPGAs [57, 59]. A similar SRAM PUF from Intrinsic-ID Inc. is implemented on Intel/Altera Stratix 10 SoCs and FPGAs [46]. Recently, Verayo Inc., a delay-based PUF developer company, announced that it will provide Xilinx Ultrascale+ SoCs and FPGAs with their PUF technology [90]. Since Xilinx has patented previously a key generation technique based on hard RO PUFs [87], most probably the PUF will be an RO PUF variant. Currently, the Xilinx Zynq-7000 SoCs enables the user to implement soft PUF IP cores as well as DPA-resistant soft decryptor IPs to protect the red key during configuration [67]. Furthermore, selected Microsemi flash-based SmartFusion2 and IGLOO2 FPGAs can be utilized as a Root of Trust (RoT) to transfer soft PUF IP cores to target SRAM-based FPGAs for secure authentication [47]. Soft PUFs can be purchased from third-party developers, such as Verayo Inc. [4], Intrisic-ID Inc. [3], and Helion Technology Limited [2].

2.3 Semi-Invasive Attacks

2.3.1 Photonic Emission Analysis

Complementary metal-oxide semiconductor (CMOS) ICs consist of individual n-type and p-type metal-oxide semiconductor (MOS) transistors connected to realize different logic gates. In any logical state of such a CMOS gate, the static current consumption is minimized because at least one transistor on the path between VDD and GND is in off region. Only during switching events, transistors pass through saturation for a short period. In saturation, MOS transistors emit photons due to carriers traveling through the space charge region near the drain diffusion [13]. The emission intensity depends on the applied voltages, conducted current, and time spent in saturation. Due to the different characteristics of holes and electrons, n-type transistors emit significantly more photons at the same conditions in comparison to p-type transistors. Hence, only changes in the logic state of n-type transistors will usually be visible during PEM.

In modern IC designs, multiple interconnect layers obstruct the optical path from the transistors drain region to the surface of the device. Hence, it is nearly impossible to observe photon emissions from the frontside of the chip. Nevertheless, photon emissions can also be observed through the silicon substrate of the IC backside. However, silicon is highly absorptive for the photons with higher energy than the bandgap energy, and therefore, near infrared (NIR) photons only will remain for analysis.

2.3.2 Laser Fault Injection

Generating electron-hole pairs by passing a laser beam through silicon is called the photoelectric effect. This phenomenon is occurred only by photons, whose energies are greater than the silicon band-gap energy. However, similar to other optical attacks, the LFI attacks are preferred to be carried out from the IC backside through the silicon substrate. Hence, the laser wavelength must be chosen from the NIR spectrum.

Depending on the laser power and its spot location on the IC, the generated electron-hole pairs might be recombined and leave only a negligible effect on the behavior of the target IC [69]. However, if the right location and power are selected, they can cause a Single Event Effects (SEE) and create an electrical current [69]. In this fashion, having a laser shot at specific regions of a SRAM cell can flip the stored value in it.

An attacker can use this principle to attack the SRAM-based LUTs of CPLDs and FPGAs. Inside the reconfigurable hardware the n-input SRAM-based LUTs store 2^n binary configuration values to realize n-input combinational logic functions, see Fig. 2.4. Thus, one can implement $2^{(2^n)}$ different combinational logic functions within

Fig. 2.6 Fault injection into LUT configuration

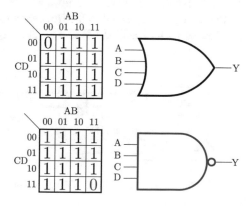

each LUT. It is clear that any changes in the state of the configuration bits of a LUT lead to a different logic function. An attacker can alter the logical function of a gate by inducing a fault into one or more SRAM cells of a target LUT. For instance, a 4-input OR gate is realized by storing 0xFFFE in the SRAM cells of a LUT [44], see Fig. 2.6. Inducing a fault into the cells 0 and 15 of the SRAM results in obtaining a NAND gate.

2.3.3   Optical Contactless Probing

Laser Voltage Probing (LVP) and Laser Voltage Imaging (LVI) techniques have been introduced in the field of failure analysis to debug the nanoscale transistors from the backside of the chip in a contactless way. Therefore, they are referred to optical contactless probing techniques. While LVP can be used to directly probe the electrical signals on the transistors, LVI can be employed to create an activity map of active circuits. In both cases, the laser photons with NIR wavelengths pass through the silicon substrate from the IC backside to reach the transistors, which leads to a partial absorption and a partial reflection of the laser beam. In the case of LVP, the reflected light is modulated based on the electrical signal on a node, and it can be fed to an optical detector to measure its intensity, see Fig. 2.7. In this way, the data passing through a node can be probed. Since the light modulation is insignificant, the signal should be measured several times and averaged by running the device in a triggered loop. In this case, a sufficient signal to noise ratio can be achieved.

For performing the LVI, on the other hand, the reflected light is fed to a spectrum analyzer with a narrow band frequency filter while the laser scans the device. In this case, the detector signal is not averaged. The laser beam is scanned across the Device under Test (DUT) using galvanometric x/y mirrors, and the filter output of the spectrum analyzer is sampled for every scanned pixel. Subsequently, a control PC is used to assemble the sampled frequency filter values into a 2D image using a grey-scale representation. If an electrical node operates at the frequency of interest, it will modulate the light reflected with the same frequency, which will be able to pass through the frequency filtering spectrum analyzer. As a result, the nodes with a switching frequency equal to the frequency filter show up as white spots in the LVI image leading localization of them on the chip.

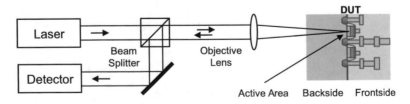

Fig. 2.7 Simplified illustration of LVP signal acquisition

2.4 Related Work

Although *unclonability* and *unpredictability* are the essential requirements for the PUFs [8], previous work in the literature has demonstrated how different PUFs can be attacked and cloned. Bistable PUFs, such as SRAM PUFs, can be read out by photonic emission analysis and physically cloned by a Focused Ion Beam (FIB) circuit edit [34]. Moreover, it has been shown that SRAM PUFs can be read out by laser stimulation [64]. Besides, SRAM PUFs are also vulnerable to remanence decay in volatile memories [65]. Finally, the vulnerabilities of the SRAM PUFs in general as a replacement for non-volatile memory are reviewed in [35].

In contrast to settling-state-based PUFs, delay-based PUFs (e.g., Arbiter PUF and RO PUF families) are believed to be resistant to physical cloning, due to their more complex and interconnected structures. The main assumption of timing-based PUFs is that only fully-invasive techniques enable an attacker to measure the individual delays within the PUF structure. This kind of attacks might alter the physical properties of the silicon substrate, which leads to undesirable changes in the CRP behavior of the PUF. However, they can still be characterized by various means. For instance, RO PUFs are vulnerable to the EM side-channel attacks [53–55] and modeling attack [24]. Arbiter PUFs have been a target for mathematical modeling attacks. Thus, the known attacks in the literature try to simulate the CRP behavior of the PUF and build a mathematical clone of it. Modeling attacks require a subset of CRPs to create a model on that and predict the PUF responses for all possible challenges [41]. It has been reported that an Arbiter PUF under the Deterministic Finite Automata (DFA) representation can be Probably Approximately Correct (PAC) learned with a given level of accuracy and confidence [26].

The modeling attacks become more difficult by introducing non-linearities to the PUF delays and responses. Two example of non-linear PUFs are Feed-forward Arbiter PUFs [42] and XOR Arbiter PUFs [79]. However, a successful modeling attack on XOR Arbiter PUFs with a limited number of Arbiter chains using logistic regression (LR) algorithm is reported [70]. In another attempt, by PAC learning the XOR Arbiter PUF with the Perceptron algorithm, a theoretical limit as a function of the number of PUF stages and the number of chains for pure modeling attack could be found [25]. Although pure modeling attacks fail to learn larger XOR Arbiter PUFs, a combined modeling attack based on a higher number of CRPs with timing and power side channel information can successfully break XOR Arbiter PUFs up to 14 Arbiter chains [72]. In another approach, the noise in the response of the Arbiter PUF was exploited as a side channel information to model the CRP behavior of the single Arbiter PUF [21]. The idea of using noise as a helper information to enhance modeling attacks is further developed by changing the temperature [22] and supply voltage of the chip [11] to induce more noise in the PUF responses. Furthermore, it has been shown that individual chain of an XOR Arbiter PUF can be separately learned by using the noise information in the CMA-ES algorithm [10].

Chapter 3
Experimental Setup

In this chapter, we present the deployed reconfigurable hardware and PoC PUF implementations on them for our experiments. Afterward, we introduce the electrical and optical setups used for PEM, LFI, and LVP/LVI. All presented experimental setups in this chapter are previously presented in [45, 81–85].

3.1 Devices Under Test

3.1.1 Intel/Altera MAX V CPLD

The first group of deployed devices under test were Intel/Altera MAX V CPLDs with part number 5M80ZT100C5N manufactured in a 180 nm process [5]. These samples were used in the PEM and LFI experiments. In this sample all Logic Elements (LE) contain 4-input LUTs and a dedicated register. The LEs in this device are arranged in groups of 10 inside a Logic Array Block (LAB). A 100 pin TQFP package was selected to simplify the sample preparation. All samples were prepared using an Ultratec ASAP-1 polishing machine. This approach allows selective die thinning of packaged samples without the need for re-bonding the samples. The samples were thinned to approximately 30 μm in substrate thickness and were inversely soldered to a custom PCB, see Fig. 3.1a. Altera MAX V CPLD contains an internal NVM to store the configuration. A JTAG connection was used to write the configuration into the NVM.

© Springer International Publishing AG, part of Springer Nature 2019
S. Tajik, *On the Physical Security of Physically Unclonable Functions*, T-Labs
Series in Telecommunication Services, https://doi.org/10.1007/978-3-319-75820-6_3

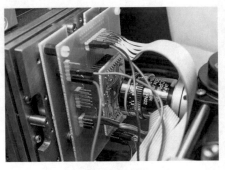

(a) Altera MAX V   (b) Altera Cyclone IV

Fig. 3.1 Devices under test: **a** A MAX V CPLD in 180 nm process manufactured by Altera. **b** A Cyclone IV FPGA in 60 nm process manufactured by Altera

3.1.2 Intel/Altera Cyclone IV FPGA

The second groups of deployed devices under test were Intel/Altera Cyclone IV FPGAs with part number EP4CE6E22C8N manufactured in a 60 nm process [6]. We chose the Cyclone IV since it is similar in architecture to the MAX V. This enables us to utilize the same PUF implementations, which allows us a direct comparison. These samples were used in the PEM and LVP/LVI experiments.

In this sample, all LEs contain 4-input LUTs and a dedicated register. The device contains 6272 LAB with 16 LEs each. We chose the 144 pin TQFP package in order to simplify the sample preparation. The first step of preparation was the removal of the exposed ground pad on the backside of the package. The samples were then thinned by an Ultratec ASAP-1 polishing machine to a remaining silicon thickness of 25 μm. In the second step, the prepared samples were inversely soldered to a custom PCB, see Fig. 3.1b. Bond wires originally leading to the exposed ground pad were then reconnected using silver conductive paint. Since Intel/Altera Cyclone IV is an SRAM-based FPGA, no internal NVM in the chip package is available. Hence, a JTAG connection was used for configuring the FPGA after each reboot.

3.2 Hardware Implementations

3.2.1 Standalone PUF Implementations

For the PEM and LFI experiments in Chaps. 4 and 5, standalone PUFs, namely Arbiter PUFs and RO PUFs, have been realized on the CPLDs and FPGAs. In the case of Arbiter PUF, each stage can be realized by two digital multiplexers on the chip. Although this design can be implemented in an optimal way on an ASIC, it

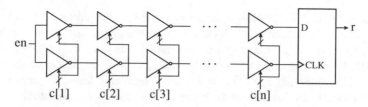

Fig. 3.2 Implementation of an Arbiter PUF by two independent inverter chains. Each challenge bit is connected to all *don't-care* inputs of the utilized LUT

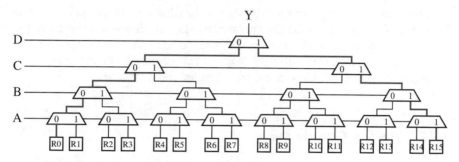

Fig. 3.3 A LUT is realized by multiple multiplexers, which are controlled by the data inputs. The output of the LUT is loaded from the existing SRAM cells inside the LUT. By connecting *don't-care* inputs A, B and C to a single bit challenge and connecting the input D to the output of previous stage, only two routes can be selected based on the challenge value

causes delay imbalances for upper and lower PUF chains on an FPGA. Moreover, the effect of the routing between the LUTs on the PUF response is more influential than the effect of intrinsic delay differences on the individual devices [63]. A better Arbiter PUF design, which is more suited for FPGAs, can be realized by two independent buffer or inverter chains as proposed in [51], see Fig. 3.2. In this case, each stage requires two LUTs. One input of each LUT is configured as the input for the incoming signal from the previous stage and all other *don't-care* inputs are connected together to a single challenge bit, see Fig. 3.3. As a result, by applying a challenge bit, two different routes inside the LUT can be involved. Each stage is placed manually in a way to make the total timing of PUF chains as symmetric as possible.

The arbiter can be implemented by dedicated registers inside the CPLD or FPGA. However, if the delay differences between upper and lower chains are less than the precision of the register, the arbiter will be in a meta-stable condition. Moreover, the propagation delays of data and clock lines inside the registers are different, and consequently, the delay differences of the upper and lower chains will be biased at this point. Thus, for our experiment we read out the response of the PUF directly by measuring the overall delays of both chains with the help of a Time-to-Digital converter (TDC) by either capturing photons from the transistors terminating the Arbiter PUF chain or measuring directly the arrival of the electrical signals at the outputs of the final stage.

In the case of RO PUF, ring-oscillators can be implemented as a logic circuit comprised of an odd number of LUTs, which are configured as inverters, connected in series to form a ring. The output of the last LUT is fed back to the input of the first LUT. Due to the odd number of inverters, as long as the circuit is powered up, every inverter-configured LUT continuously oscillates between 0 and 1. For the LFI experiments, ring-oscillators with five inverters plus one logical AND gate for enabling and disabling the ring have been implemented. The entire circuit of each ring-oscillator was placed within a single LAB, and each inverter was placed in an individual LE within the LAB.

Binary counters using dedicated registers of the CPLD or FPGA can measure the frequency of the individual ROs in a predefined interval. However, to have a higher precision, we measured the frequency of all ring-oscillators directly by oscilloscope and spectrum analyzer. In this case, the outputs of the ring-oscillators were connected directly to the pins of the chip for a direct electrical measurement.

3.2.2   Red Key Calculation

For the LVP and LVI experiments in Chap. 6 we have implemented an RO PUF and a red key (see Sect. 2.2.1) calculation on the FPGA. To make the design less complex, we have connected the outputs of the ROs directly to individual counters, see Fig. 3.4. Each ring-oscillator in our design has been realized with 21 inverters. All components of the ROs and the counters have been placed manually inside the FPGA using the Altera Quartus II integrated development environment. The LEs in every ring-oscillator were placed as close as possible, directly next to each other. We have emulated the rebooting and configuration of the FPGA by adding a reset signal to our implementation. The black key and PUF key in our design have 8-bit length. As discussed in Sect. 6.1, unwrapping the black key can be carried out either in a parallel or serial way. Hence, for the first scenario, we have implemented the red key generation by XORing all values of the black key with the PUF key in parallel, see Sect. 6.1.1. For the second scenario, we have realized two shift registers for the black key and PUF key, where those values are shifted serially to an XOR gate and the result is shifted into the red key registers.

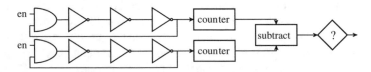

Fig. 3.4 A simplified schematic of a ring-oscillator pair in the RO PUF construction. After a predefined period of oscillation, the states of both counters are compared to each other to generate a binary response

3.2.3 PUF-Based Sensor

To design a PUF-based sensor for LVP and LVI detection in Chap. 6 we implemented a 16-bit RO sum PUF comprised of 32 ROs on the FPGA. Each RO in our design comprised of 5 inverters. The ROs were placed manually and distributed all over the FPGA silicon area, which utilized about 15% of FPGA resources. The frequencies of ROs are measured by 16-bit counters. The challenges are transmitted from a laptop via the UART protocol to the FPGA and the generated responses are sent back on the same channel to the laptop.

3.3 Optical Setup

3.3.1 Photonic Emission Analysis Setup

The experimental setup consists of an optimized infrared microscope equipped with a Si-CCD camera and an InGaAs avalanche diode as detectors for spatial and temporal analysis [75], see Fig. 3.5. The Si-CCD is cooled down to $-70\,°C$ to minimize dark current. This allows long exposure times to accumulate enough photons from the weak hot carrier emission. Since the integration time of CCD camera is several seconds and the readout speed of its sensor is limited, the Si-CCD camera is used for spatial photonic emission analyses only. Hence, a very fast infrared detector is required for the temporal photonic emission analysis. A free-running InGaAs avalanche detector in Geiger Mode (SPAD) can fulfil this requirement and be used to detect single photons. The sensitivity of SPAD covers a wavelength range between 1 to $1.6\,\mu m$. A computer controls the DUT via a control box (CB), which provides the trigger pulse, i.e., enable signal, for the implemented Arbiter PUF and a time reference signal for the FPGA-based TDC. The emitted photons from the DUT are collected by the microscope objective with an NA = 0.6 and divided into two optical paths by a short-pass beam splitter (BS). While short-wavelength photons below $1\,\mu m$ are transmitted to the Si-CCD camera, the long-wavelength photons are reflected onto the InGaAs-SPAD. This configuration allows spatial and temporal photonic emission analysis simultaneously. An incoming photon from the DUT causes the avalanche breakdown of the SPAD and the resulting electrical pulse is registered by the TDC. The TDC register the time of each occurring event with a resolution of $81\,ps$. In this case both the enable signal of the Arbiter PUF and the detected photons from the transistor terminating the chain are time tagged allowing a direct calculation of their delay. The overall time uncertainty for a single photonic event is $190\,ps$ rms. This is due to the jitter in the response time of the SPAD and electrical jitter in the CB and TDC. However, an accumulation of multiple photonic events is used to increase the time resolution by computing the mean value of the Gaussian-like distribution of the delay time histogram. This technique enhances the time resolution significantly beyond the $81\,ps$ granularity of the TDC and allows measurements of timing differences at the end of the PUF chain for two different challenges. As a

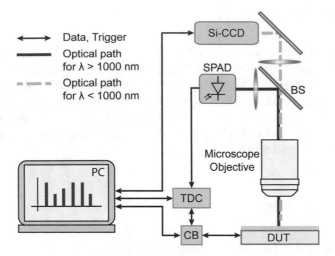

Fig. 3.5 Schematic of photonic emission setup

result, the accuracy of our time-resolved measurement setup is limited by drifts in the electronics to 6 ps rms. The setup contains a custom made holding of the DUT to a 3-dimensional moving stage and electronics to control and communicate with the DUT.

3.3.2 Laser Fault Injection Setup

A Hamamatsu PHEMOS 1000 photon emission and laser scanning microscope was used for LFI experiments conducted in this work (see Chap. 5). The system is a commercially available instrument for failure analysis of semiconductor devices. A silicon CCD sensor is used to capture the photonic emission of transistors. The sensor is cooled down thermo-electrically to −50 °C to minimize dark current and noise.

There are two different laser sources with the wavelengths 1064 nm or 1319 nm available on the system. The laser beam is scanned in a zigzag pattern across the DUT, and the reflected light signal is sampled at constant intervals. The individual samples of the reflected light signal are then assembled digitally, corresponding to the laser beam position, and consequently to their physical location on the DUT. In this way, it is possible to acquire a pixel-by-pixel image of the reflected light pattern of the DUT, which can then be used for navigation. Figure 3.6 shows an acquired reflectance image of an Altera MAX V CPLD from the backside. An array of 4 by 6 blocks can be identified, corresponding to the 24 LABs each containing 10 LEs. The non-volatile memory responsible for storing the configuration data, as well as the programming circuitry and additional hard-macros available to the user, can be

Fig. 3.6 Backside
reflectance image of the
CPLDs used throughout this
work. The framed area
contains the programmable
logic cells. The grid
corresponds to the placement
of 4 by 6 LABs. Each LAB
contains 10 LEs (only shown
for one LAB)

observed in top half of the reflectance image. Along the perimeter of the device, the
I/O pads are clearly visible.

In the LFI experiments, 20x/0.4 NA Mitutoyo objective lens was utilized for
long distance navigation. For fault injection attempts and short distance navigation,
a 50x/0.76 NA Hamamatsu objective lens was used. In our LFI experiments, only
the laser with 1064 nm wavelength was deployed. This laser source can operate
high power impulse mode and low power mode. The low power mode is used for
navigation only, and the high power impulse mode is used for fault injection. In low
power mode, the maximum laser power at the laser source is 200 mW. In impulse
mode, the maximum power at the source is 2 W with a pulse duration of 200 ns. The
laser power can be adjusted from 2 to 100% in 0.5% steps.

After finding the target LEs by spatial photonic emission analysis, the electrical
setup assists us to find the sensitive locations of an LE for LFI in an automated way.
In this setup, the laser source scans across the device, and the laser is shot once for
every pixel in high power mode. Following the laser shot, the device is first examined
for changes in configuration. Afterward, the laser beam moves on to the next pixel
position to repeat the same procedure. The obtained configuration changes are then
assembled into an image which represents the response to the laser pulse across the
examined LE area.

To realize a proper electrical setup, a set of requirements has to be met. First, since we aim to reconfigure the LUTs by injecting faults into the configuration SRAM cells, we need to assure that the DUT is properly reconfigured after each laser shot. Therefore, the DUT has to rebooted for every pixel to trigger the reconfiguration of the LUT SRAM cells. Second, the configuration of DUT has to be tested in a certain time after the laser shot. This step is required because we are only interested in semi-permanent changes (i.e., the faults are permanent as long as the FPGA is powered on or not rebooted) in the device behavior rather than temporary changes, which are only present during laser irradiation. In conjunction with this, after the fault injection, we have to guarantee that the device is still functioning properly and has not entered into an entirely dysfunctional state. Based on these requirements, a setup as shown in Fig. 3.7 should be assembled.

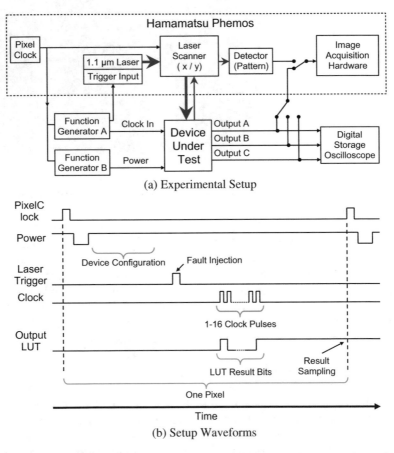

(a) Experimental Setup

(b) Setup Waveforms

Fig. 3.7 Schematic of laser fault injection setup: **a** Block diagram of our optical and electrical setup, which is utilized for the laser fault injection attack. **b** Signal flow of measurement

In this setup, PHEMOS generates a pixel clock output, which is synchronized with the scanning mirrors. This pixel clock is a primary trigger for all the other electrical signals. As soon as the pixel clock signal goes low, function generator B disconnects the power supply to the DUT for a short amount of time and then reconnects the power to the DUT, see Fig. 3.7a. The reconfiguration of CPLD (see Sect. 3.1.1) takes a maximum of $200\,\mu s$ [5]. An extra delay is added to make sure that the device has reached a stable state after reboot. The first channel of the function generator A triggers the laser shot after this delay, see Fig. 3.7b. Following that, a second delay is added to allow the device to reach a stable state after the disturbances caused by the laser. Afterward, the second channel of the function generator A supplies 16 clock pulses to a 4-bit binary counter in the DUT. The outputs of this counter have been connected to the inputs of the target LUT, and therefore, the number of clock cycles will determine which SRAM cell value is loaded at the output of the LUT. As the output of the LUT is connected to the PHEMOS pixel-value input, the output of the LUT will be sampled by the PHEMOS as soon as the pixel clock goes high. The procedure is repeated for the next pixel again. For one pixel the whole process takes about one millisecond. In this way, we acquire an image which indicates the state of the examined LUT bit after the fault injection for every position of the scanned area. Due to the delay between the laser shot and the actual sampling of the DUT result, one can be sure that the pixels are indicating a permanent state change of the examined bit. The sensitive locations for the different LUT result bits can then be examined by selecting a different number of clock cycles. In this way, all locations, which trigger a change in one of the LUT result bits, can be found. Finally, the laser can be shot at these locations manually to induce configuration changes. The general region which was identified to be sensitive to fault injection can be seen in Fig. 3.8.

For instance, we have programmed the LUT of an LE with a 0x8000 value, which is translated to an AND gate with four inputs, see Fig. 3.9a. The AND gate can be represented by $Y = A.B.C.D$, where Y is the output of the LUT and A, B, C and

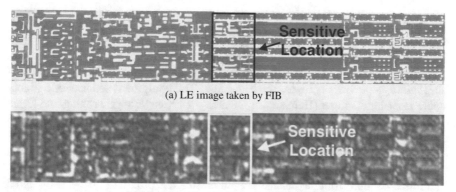

(a) LE image taken by FIB

(b) LE image taken by laser scanning microscope

Fig. 3.8 The sensitive regions of an LE to the LFI

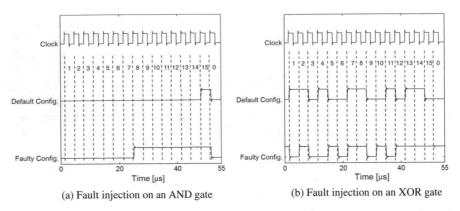

(a) Fault injection on an AND gate   (b) Fault injection on an XOR gate

Fig. 3.9 **a** LFI against AND gate. **b** LFI against XOR gate. Note that the numbers in both graphs are indicating the address of the SRAM cells

D are the four inputs. By inducing random faults into the sensitive regions of the target LE, we were able to manipulate the configuration of the LUT to 0xFF00, which can be represented by $Y = D$, see Fig. 3.9a. The new configuration accounts for a buffer for the first input of the LUT, which considers the other three inputs as *don't care*. In another attempt, we have programmed an XOR gate in the LUT with four inputs, where the stored configuration value is 0x6996, see Fig. 3.9b. The XOR function can be written as $Y = A \oplus B \oplus C \oplus D$. The laser shot converts the XOR gate to a more complex combinational function with the configuration 0xF5A5, which represents the function $Y = A.C + !A.(D.!C)$, see Fig. 3.9b. Both results demonstrate the successful manipulation of LUT configurations. However, due to the limited precision of our laser setup, the generated faulty configurations are random, and they cannot be controlled completely.

3.3.3 Optical Contactless Probing Setup

To perform optical contactless probing the Hamamatsu PHEMOS-1000 laser scanning microscope was deployed, see Fig. 3.10a. The equipment consists of a highly stable laser light source (Hamamatsu C12993), a Laser Voltage Probing and Laser Voltage Imaging preamplifier (Hamamatsu C12323), an Agilent Acqiris digitizer card and an Advantest U3851 spectrum analyzer. The laser light source emits photons with a wavelength of 1319 nm. The emitted light is deflected by galvanometric mirrors and then focussed through an objective lens into the backside of the DUT. The reflected light from the DUT is passed on to a detector, and the detector signal is fed into the preamplifier. The output of the amplifier can either be connected to the digitizer card for the acquisition of LVP waveforms or to the spectrum analyzer for LVI. For all probing experiments, a Hamamatsu 50x/0.76NA lens was used. In

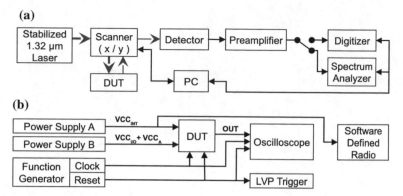

Fig. 3.10 Schematic of Laser Voltage Probing setup: optical (**a**) and electrical (**b**) setup block diagram

the case of 100% of laser power, the approximate laser power with this lens on the DUT is 50 mW. 5x and 20x objective lenses were used only for navigation purposes. A PC running the PHEMOS software controls the optical setup.

The electrical setup consists of two power supplies, which are connected to the DUT, see Fig. 3.10b. The first power supply (Agilent E3645A) supplies the internal logic of the FPGA with $V_{CCINT} = 1.2$ V. The second power supply (Power Designs Inc. 2005) supplies the I/Os and the analog circuits of the FPGA (see Sect. 3.1.2) with $V_{CCIO} = 2.5$ V (I/O) and $V_{CCA} = 2.5$ V, respectively. A two channel function generator (Rigol DG4162) generates clock and reset signals which are fed into the DUT. The clock signal, reset signal, and an auxiliary DUT output are connected to an oscilloscope (LeCroy WaveMaster 8620 A) at the same time for testing and control purposes. The reset signal is furthermore fed into the LVP trigger input. To perform basic power analysis in the frequency domain, a Software Defined Radio (SDR) is AC-coupled to the V_{CCINT} power rail. The SDR is an USB dongle which utilizes a Realtek RTL2832U chipset and a Rafael Micro R820T tuner.

Chapter 4
Photonic Side-Channel Analysis

In this chapter, we demonstrate that the primary security assumption on the infeasibility of direct delay measurements in delay-based PUFs is not valid. By performing an experiment on a PoC Arbiter PUF implementation on reconfigurable hardware, it becomes apparent that The Arbiter PUF family and more generally, the delay-based PUFs can be characterized by a high-resolution temporal photonic emission analysis. This approach does require neither any access to the PUF's response nor a significant number of challenges to characterizing the PUF. This chapter with slight revisions are based on publications [81, 82].

4.1 Attack Scenario

As discussed in Sect. 2.1.1, infeasibility of the direct delay measurement of each stage in an Arbiter PUF is the main security assumptions of these primitives. If the attacker tries to measure the delays of each stage using invasive microprobing, the intrinsic internal delays of the PUF is changed with a high probability [74]. In other words, the challenge-response behavior of the PUF is altered, and therefore, the PUF will be destroyed.

However, it is known that side-channel analysis of the Arbiter PUF can reveal the total signal propagation delays at the outputs of the last stage [11, 72]. For instance, the attacker can measure the delay differences between the PUF activation time and the logical transition time of the Arbiter by conducting the power analysis [11, 72]. By gathering several power traces for a single challenge and applying statistical signal processing techniques on them to reduce the noise of measurements, the total delays of the chain and the response of the PUF can be extracted. The main limitation of power analysis is that the delay differences of upper and lower chains of the PUF for a given challenge cannot be derived from the power traces.

Timing side-channel analysis is another option to measure the delays of both upper and lower chains individually [72]. To collect the timing information, the attacker has to have access to the dedicated debugging clock sweeping circuits on the chip,

© Springer International Publishing AG, part of Springer Nature 2019

S. Tajik, *On the Physical Security of Physically Unclonable Functions*, T-Labs
Series in Telecommunication Services, https://doi.org/10.1007/978-3-319-75820-6_4

which is embodied in the chip by the manufacturer and used to test the quality of the implemented PUF [51]. By changing the frequency of the clock, which is swept through each chain of an Arbiter PUF, the attacker can measure the delays of each path. Although timing side-channel analysis enables the attacker to measure the total delays of both chains with a picosecond resolution, the assumption that the attacker can access the debugging circuit might not be valid in a real scenario.

To measure the delays of the chains without using any extra debugging circuitry, the attacker can deploy high-resolution temporal photonic emission analysis from the IC backside. In this case, the attacker can measure the delay between the activation time of the PUF and the emission time of photons from a CMOS transistor at the outputs of *any arbitrary* stage by using an Avalanche Photodiode (APD). Figure 4.1 shows the timing differences of emitted photons at one of the outputs of the last stage of the Arbiter PUF by applying two different challenges. This measurement can be performed using the experimental setup introduced in Sect. 3.3.1.

4.1.1 Characterization of Multiplexer-Based Arbiter PUF

In a multiplexer-based Arbiter PUF, each stage consists of two direct paths and two crossed paths, see Fig. 2.1. To completely characterize an n-stage Arbiter PUF, the propagation delays of each path have to be known, and hence, $4n$ delays must be characterized in total. One way would be to naively measure all 4 propagation delays at all n stages individually by moving the optical setup *over* both inputs and both

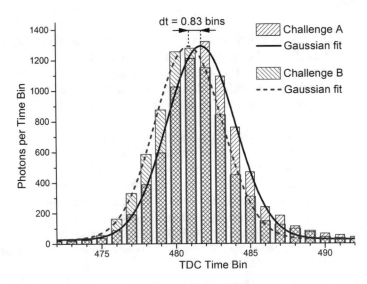

Fig. 4.1 Timing difference of two different challenges at the output of last stage. The time bin width is 81 ps

outputs of each stage, and only try both challenge states. However, this technique would require the movement of the optical objective and adjust the focus for each movement. In this case, a precise aperture movement can be automated and eventually yield the $4n$ Arbiter delays. A more optimal solution can be the measurement of the overall propagation delays of each PUF chain at the outputs of the very last stage for sufficiently many selected challenge combinations. Due to the additive linear model of the Arbiter PUF [42], the overall delay at the outputs of the last stage is the sum of all n delays in each stage. Therefore, every measurement has to consider for every chosen challenge the complete propagation time of two distinct but possible paths. If we denote by r_i the resulting overall time of an individual challenge measurement, we conclude that we get an inhomogeneous system of linear equations

$$\mathbf{C} \cdot \delta = \mathbf{r}$$

for our $4n$ unknowns $\delta_{i_a}, \delta_{i_b}, \delta_{i_c}$, and δ_{i_d} and the challenge matrix $\mathbf{C}$ with entries from $\{0, 1\}$ which encode the different valid paths through the Arbiter chain. We call a path $\mathbf{c}_i \in \{0, 1\}^{4n}$ *valid* if its respective challenge setting within $\mathbf{C}$ allows a full signal propagation of length n. By induction the following is easy to observe.

Proposition 1 *For an Arbiter PUF of length $n \geq 1$ let $\mathbf{C}$ be the $(2^{n+1}) \times (4n)$ matrix consisting of all valid paths through the respective Arbiter chain. Then* $\mathrm{rk}(\mathbf{C}) = 2n + 2$.

Having only $2n + 2$ linear independent equations in $\mathbf{C}$, we need to generate the remaining $2(n - 1)$ linear independent equations in another way to completely solve our system. Thus, instead of full propagation paths we are forced to consider partial valid paths as well. Let $\mathbf{c}_i \in \{0, 1\}^{4n}$ be a valid path; for integers $1 \leq u, v \leq n$ a vector of the form

$$(0, \ldots, 0, c_{4u}, c_{4u+1}, c_{4u+2}, c_{4u+3}, \ldots, c_{4v}, c_{4v+1}, c_{4v+2}, c_{4v+3}, 0, \ldots, 0) \in \{0, 1\}^{4n}$$

will be called a *partial valid* path.

Note 1 *For a partial valid path we will measure its signal time only from the inputs of Arbiter stage u until its output at stage v and deliberately denote this partial time simply also by r_i.*

Including such partial measurements r_i and their corresponding paths $\mathbf{c}_i$ we get by induction.

Proposition 2 *For an Arbiter PUF of length $n \geq 1$ and its $2n + 2$ valid paths (corresponding to the linear independent row vectors) there exist $2(n - 1)$ appropriate partial valid paths such that their combined challenge matrix $\mathbf{C}$ has full rank $4n$.*

This Proposition implies that we only require $2(n - 1)$ partial measurements which we classify with respect to u and v into three classes:

1. $u = 1$ and $1 \leq v < n$: Measurement begins at the inputs of the first stage and ends in the middle of the chain.

2. $1 < u, v < n$: Measurement starts at some inputs in the middle of the chain and
   also ends in the middle of the chain.
3. $1 < u \le n$ and $v = n$: Measurement starts at the inputs in the middle of the
   Arbiter chain and and ends after the last stage.

To keep the previously discussed physical measurement efforts minimal, it is therefore obvious to generate the missing linear independent equations out of group 1
or 3. This completes our description of an optimized measurement for a classical
multiplexer-based PUF with n stages.

4.1.2 Characterization of Inverter-Based Arbiter PUF

In the case of the inverter-based PUF, the upper and the lower path are not crossing
at all. Hence, we can consider them entirely separately, see Fig. 3.2. Towards this,
let us consider the upper path and simply denote its n unknown delays by $\delta_1, \ldots, \delta_n$.
In other words, setting the respective ith challenge bit to 1 adds the delay δ_i to the
overall complete signal propagation time which will be denoted by r_j for the jth
measurement from the first input until the last output—just through all n stages.
If we now define the distinguished variable Δ_{n+1} as the overall complete signal
propagation time for setting all n challenge bits to 0 we get the linear system

$$
\begin{pmatrix}
1 & 0 & \cdots & 0 & 0 \\
0 & 1 & \cdots & 0 & 0 \\
\vdots & & \ddots & \vdots & \vdots \\
0 & 0 & \cdots & 1 & 0 \\
0 & 0 & \cdots & 0 & 1
\end{pmatrix}
\cdot
\begin{pmatrix}
\Delta_1 \\
\Delta_2 \\
\vdots \\
\Delta_n \\
\Delta_{n+1}
\end{pmatrix}
=
\begin{pmatrix}
r_1 \\
r_2 \\
\vdots \\
r_n \\
r_{n+1}
\end{pmatrix}
$$

for which we simply require the measurements $r_i, i = 1, \ldots, n + 1$. The lower path
can be handled in an analog way, say $\mathbf{C}' \cdot \mathbf{\Delta}' = \mathbf{r}'$. Moreover, using the unit vectors
$\mathbf{e}_i \in \{0, 1\}^{n+1}, i = 1, \ldots, n + 1$, we find that we get from

$$\mathbf{e}_i \cdot \mathbf{\Delta} - \mathbf{e}_{n+1} \cdot \mathbf{\Delta} = r_i - r_{n+1}, \quad \text{and}$$

$$\mathbf{e}_i \cdot \mathbf{\Delta}' - \mathbf{e}_{n+1} \cdot \mathbf{\Delta}' = r_i' - r_{n+1}'$$

the two individual inverter delays δ_i and δ_i' of stage i incurred by setting the ith challenge bit to 1. We thus conclude that we need only $2n + 2$ "full path" measurements
to completely characterize a delay-based PUF with n stages.

4.2 Results

A PoC Arbiter PUF is implemented on the Altera MAX V CPLD (see Sect. 3.1.1) with 8 stages. We selected the challenge 00000000 as the reference challenge for our characterization measurements. To measure the effect of each challenge bit, we applied the challenge combinations with Hamming distances one to observe the effect of each challenge bit individually. The chip was supplied with 2.2 V, and the enable signal was switched with a frequency of 4 MHz. The photonic emission of the composing transistors of the PUF circuit reveals the position of each stage, see Fig. 4.2.

If the electrical access to the challenges is restricted, comparing the obtained spatial emission images of the PUF stages can also reveal the state of individual challenge bits. By changing each challenge bit, the emission pattern of each LE is changed, and therefore, the challenge can be read out without any electrical access to it, see Fig. 4.3. Therefore, the equations provided in Sect. 4.1.2 can still be used to characterize the PUF by finding challenges with Hamming distance one from each other.

After setting a challenge, the PUF was activated 50 million cycles to capture enough number of photons for analysis. The reference challenge also has been measured several times during our experiments to compare the consistency of measurements. The measurement results of 8 challenge combinations compared to the reference challenge can be found in Table 4.1. Positive timing difference means that the delay is decreased in comparison to the reference challenge and vice versa. It

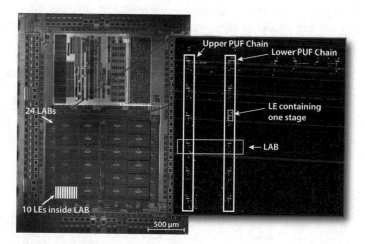

Fig. 4.2 The backside reflectance image acquired using a laser scan microscope (left). Optical emission of the 8-bit Arbiter PUF on the CPLD (right). Each stage is realized by two LEs in a LAB in parallel

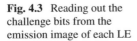

Fig. 4.3 Reading out the challenge bits from the emission image of each LE

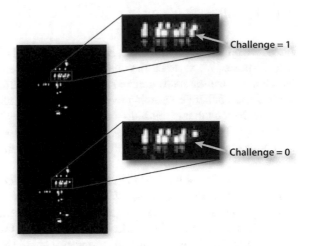

Challenge = 1

Challenge = 0

can be seen that flipping the challenge bit from 0 to 1, makes in all cases both upper and lower chains faster. Moreover, the timing differences between both chains can also be found in Table 4.1. Based on the overall delay difference of two chains, the response of PUF can be predicted. In this case, if the timing difference between two chains is positive, the response is 1. Otherwise, the response is 0. If there is no timing difference between the chains, the response is undefined.

According to the measured values, we can predict the behavior of both chains for all other challenge combinations based on the linear additive model of the Arbiter PUF. To prove the applicability of this model, we predicted the overall delay of both chains for a set of arbitrary challenge combinations theoretically, and then measured

Table 4.1 The CPLD optical measurement results of challenge combinations with hamming distance one (the 8 combinations from the left). Measurement results of set of arbitrary challenge combination (the last 8 combinations from the right). The reference challenge is 00000000

		1	0	0	0	0	0	0	0	0	0	1	0	0	1	0	1
		0	1	0	0	0	0	0	0	0	0	0	0	1	1	1	1
		0	0	1	0	0	0	0	0	0	0	1	0	0	1	1	1
Challenge		0	0	0	1	0	0	0	0	0	1	0	0	1	0	0	1
		0	0	0	0	1	0	0	0	0	0	1	0	0	0	0	1
		0	0	0	0	0	1	0	0	1	1	0	1	1	0	1	1
		0	0	0	0	0	0	1	0	1	1	0	0	0	1	1	1
		0	0	0	0	0	0	0	1	1	0	0	0	1	0	1	1
measured Δt in ps	chain u	48	75	38	68	49	81	49	88	217	147	184	130	312	164	355	500
	chain l	33	78	43	90	42	80	40	88	207	169	159	119	335	155	330	495
	diff.	15	-3	-5	-22	7	1	9	0	10	-22	25	11	-23	9	25	5
measured response		1	0	0	0	1	1	1	x	1	0	1	1	0	1	1	1
calculated Δt in ps	chain u									218	149	184	130	312	161	331	496
	chain l									208	170	158	120	336	154	329	494
	diff.									10	-21	26	10	-24	7	2	2
calculated response										1	0	1	1	0	1	1	1

the timings in practice. For instance, the calculated timing difference between both chains for the challenge 00000111 is the sum of measured differences of challenges 00000001, 00000010 and 00000100, which is 7 ps. The measured value is 9 ps, with 2 ps deviation from the predicted value. This example shows that the prediction is accurate enough for this specific challenge. However, it still cannot guarantee the same accuracy for all other possible combinations. Hence, we have measured the delays at the end of both chains for all 255 possible challenges. As it can be found in Fig. 4.4, the circle dots are showing the optically measured propagation delay differences of both chains from challenge 1 (i.e., 10000000 in the binary format) to challenge 255 (i.e., 11111111 in the binary format). Note that the challenge 0 (i.e., 00000000 in the binary format) is the reference challenge. The square dots show the deviation of the predicted values from the real values, see Fig. 4.4. As it can be seen, the average of the deviation between the measured and predicted values are meaningfully less than typical delay differences at the end of the chain, which does not affect the response prediction. To calculate the precise error rate of response prediction, we compared all predicted responses with the real responses based on optical measurements, see Fig. 4.5. Out of 247 applied challenges, the responses of 12 challenges are predicted incorrectly. Hence, we could obtain the prediction accuracy of 95%.

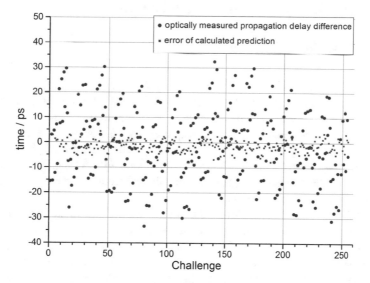

Fig. 4.4 Optically measured propagation delay differences on the CPLD and error of calculated prediction for all possible 255 challenges in picoseconds

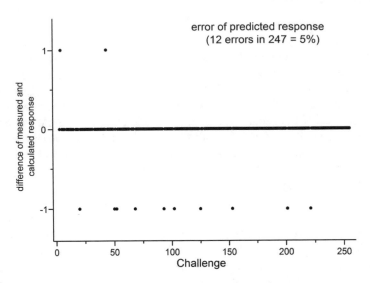

Fig. 4.5 Error of predicted responses for all possible 255 challenges. The error is rate is around 5%

4.3   Discussion and Potential Countermeasures

4.3.1   Feasibility of the Attack

Measuring the effect of each challenge takes approximately 12.5 s when supplying the chip with 2.2 V and enabling the PUF input with 4 MHz frequency. Supplying the chip with 1.8 V, for example, reduces the number of emitted photons by a factor of 3, and the measurement time increases consequently by a factor of 3. However, we can increase the frequency to 100 MHz to increase the number of emitted photons and to reduce the measurement time. Furthermore, immersion objectives or objective lenses with a larger numerical aperture (NA) can be utilized to reduce the measurement time to less than 1 s for each challenge.

Physical characterization of Arbiter PUF assists the attacker to predict the response to any arbitrarily applied challenge. Prediction of responses for unseen challenges enables the attacker to create a CRP Lookup table in the software or hardware to emulate the CRP behavior of the Arbiter PUF, which is referred to a digital clone. Besides, measurement of the rigorous delays might enable the attacker to create even a physical clone of the Arbiter PUF. To this end, it is possible to have an accurate delay map of the LEs on a second platform and try to utilize those, whose delays are close to the stages of the target PUF. Another option, though it is much more expensive, is to edit the circuit delays of the second platform with the help of FIB [74] to obtain timings close to timings of the target PUF. Thus, although achieving a physical clone of an Arbiter PUF is an onerous task, it is in principle possible.

4.3.2   Photonic-Side Channel Attack Versus Modeling Attack

We have to consider two different scenarios to compare our proposed side-channel attack with modeling attacks. In the first scenario there is no mechanism hiding the challenges and responses, and therefore, the attacker has direct access to the CRPs of the PUF. In the second scenario, a non-linear architecture of PUFs, such as XOR Arbiter PUFs [79], hide the response of each Arbiter chain from the attacker. Furthermore, the responses can be permuted or transformed by employing controlled mechanisms such as hash functions [27].

In the first scenario, the modeling attacks can be very efficient in practice, since the number of CRPs required to retrieve the response for an unseen challenge is not enormous [70]. The main advantage of the modeling attacks over photonic side-channel analysis is that they are much more cost-effective. Moreover, the semi-invasive attacks require direct physical access to the DUT, while it might not be the case for the modeling attacks. However, in the second scenario, where the responses of the multiple Arbiter PUFs are XORed, the effectiveness of modeling attacks is impaired. It has been *proved* that the pure modeling techniques can break the security of XOR Arbiter PUFs only with a limited number of Arbiter chains [25]. Although combining modeling attacks with side-channel information can relax this theoretical restriction, there still exists a bound on the effectiveness of these attacks [72]. Furthermore, the modeling attacks could be theoretically very weak when the response of the PUF is not available.

The strength of photonic side-channel is revealed in the second scenario, where no electrical access to the responses is available. As our proposed attack measures the delays of PUF chains directly before the Arbiter, obtaining the generated responses is irrelevant. Therefore, an XOR Arbiter PUF can be fully characterized regardless of the number of Arbiter chains. It is obvious that the number of required challenges in our approach increases only linearly when growing number of stages. In a similar way, each and every controlled mechanism on the response of the PUF can be bypassed. Moreover, even when the challenges are controlled, e.g., by performing a hash function, a lattice basis reduction attack can be launched [23]. In this case, the measured delays via photonic emission and the number of stages of the Arbiter PUF are the only inputs required to disclose the hidden challenges, and finally, the individual delays of each stage.

4.3.3   Applicability of the Attack on Smaller Technologies

The emission intensity is reduced by the chips with smaller technologies, due to their lower supply voltage. Moreover, the reduced distances between the shrunken transistors might prevent the attacker to distinguish the transistors from each other by conducting spatial photonic emission analysis, and therefore, the PUF stages cannot be located. The question then arises whether the same spatial photonic emission analysis of Arbiter PUFs can be applied on the chips relying on smaller technologies. To answer this question, we have utilized the Altera Cyclone IV FPGA (see Sect. 3.1.2) manufactured with 60 nm process [6] as a DUT. The FPGA was supplied by 1.4 V and the enabling signal was switched with a frequency of 4 MHz As the feature size of Cyclone IV is three times smaller than of MAX V, it is expected that the corresponding downscaled size of the transistors results in an intense decrease of the photon emission rate. A comparison of photon emission images of both ICs is shown in Fig. 4.6. Despite the small feature size of 60 nm adjacent LEs in the Cyclone IV are clearly resolved as well as parts of their inner structure. However, the photon emission rate of the relevant transistors is about ten times lower in Cyclone IV as compared to MAX V, which at least increase the required measurement effort tenfold. In order to estimate the necessary effort, we started with electrical measurements of the propagation delays for each challenge by connecting the electrical output of the last stage of every PUF chain directly to the TDC, see Fig. 3.5. A timing accuracy of 0.5 ps is achieved in the electrical measurements.

Figure 4.7 shows the propagation delays of both Arbiter PUF chains for each challenge with regard to the propagation delays of reference challenge 0 in decimal representation. As can be seen, every stage of chain 2 contributes to a delay difference of about 5–20 ps to the delay of the chain, which is resolvable by optical measurements. Whereas in chain 1 only 2 of the 8 stages showed a challenge dependency, which is insufficient for our analysis. Hence, we compared the timings of many LEs of the Cyclone IV to realize a different chain 1 path that has more challenge dependent stages. The analysis of all LEs revealed that the variance of their propagation delays, except a few, is too small for this type of Arbiter PUF implementation. As Fig. 4.7 shows, the derived response of the PUF is dominated by chain 1. Further optical measurements on such a system are pointless until a better implementation of

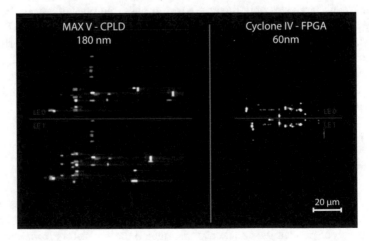

Fig. 4.6 Emission image of two inverters of one PUF stage in neighboring LEs on 180 and 60 nm. Both LEs are clearly identifiable in the image. Note that the shown LEs on the FPGA are mirrored horizontally

the PUF stages on the FPGAs is found. However, our experiments proved that photon emission still can be used to assess the signal propagation and structural properties of ICs with feature size down to 60 nm and is, therefore, a powerful tool for a physical characterization of Arbiter PUFs.

4.3.4 Countermeasures

Launching a successful photonic SCA against Arbiter PUF has two requirements. To get enough photons for accurate measurement of the delays several reactivations of the PUF with the same challenge is required. Moreover, to characterize Arbiter PUF in an optimized way, the attacker needs to apply challenge combinations with Hamming distances one. A potential countermeasure could be an algorithmic procedure independent of the PUF, which prevents the attacker from reactivating the PUF with the same challenge and applying arbitrarily chosen challenges to get low-pairwise Hamming distances.

In an authentication scenario, the availability of the adaptive chosen challenges and repeated measurements can be limited by a proposed server managed CRP lockdown protocol [95]. While the lockdown protocol is primarily designed to prevent repeated measurements in the case of ML attacks with noise-side channel information [10], it is effective against the photonic SCA as well. In this protocol, a set of CRPs is used to authenticate a device to a server, where neither the device nor the server can alone determine all challenges of the set. In this setup, two identical Linear Feedback Shift Registers (LFSR) are required to be deployed both on the device and server as Pseudo-Random Number Generators (PRNG). Furthermore, A True Random Number Generator (TRNG) is needed on the device to generate a nonce. Finally, a software-based authentication verification model of the Arbiter PUF is stored on the server.

In the first communication phase, a set of challenges, which are called device challenges, is generated on the device by the TRNG and sent to the server. Upon receiving the device challenges, the server generates a second set of challenges, which are called server challenges. Both challenge sets are concatenated and fed into the PRNG on the server to determine the ultimate set of challenges. Afterward, this ultimate challenge set is applied to the software-based verification model of the PUF and a set of responses are generated. In the second communication phase, the server challenges and a first part of the responses are sent to the device. Subsequently, the device can determine the same ultimate set of challenges by concatenating device challenges and server challenges to one set and feeding it into the PRNG. By applying the ultimate set of challenges to the PUF on the device, a response set is obtained. In this case, fractional Hamming distance of the first part of generated responses and received responses from the server can be checked on the device. If the fractional Hamming distance is above a threshold, the authentication is aborted. Otherwise, in the final step of the communication, the second part of the responses are sent by the device to the server. The fractional Hamming distance of received responses

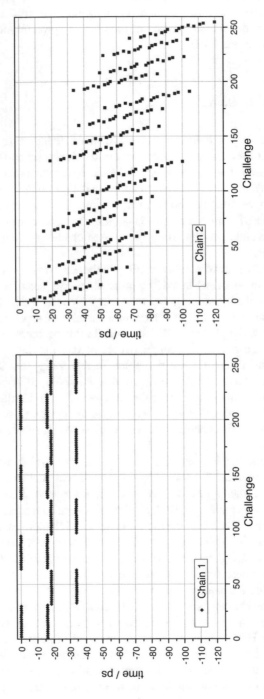

Fig. 4.7 Electrically measured signal delays of both Arbiter PUF chains in a Cyclone IV

checked against the second part of the responses, which are previously generated on the server. In a similar fashion, if the fractional Hamming distance is above a threshold, the authentication is aborted. Otherwise, the authentication is successful.

By limiting the number of possible authentication attempts, the chance of obtaining an identical ultimate challenge set form the PRNG is very low. Moreover, upon each restart, the TRNG generates a random device challenge, which is not predictable and cannot be deployed for repeated measurements. Therefore, the attacker cannot reactivate the PUF with the same challenge and apply arbitrarily chosen challenges to get low-pairwise hamming distances, and hence, the success probability of the photonic SCA is drastically reduced.

Chapter 5
Laser Fault Injection

In this chapter, we demonstrate the vulnerabilities of the soft PUF implementations on the reconfigurable hardware against LFI attacks. The building blocks of a soft PUF implementation are realized by identical programmable logic cells. It is evident that any faults in the configuration memory of deployed logic cells change the logical functionality of that cell, and consequently, could affect the PUF behavior. We present an LFI attack against PoC XOR Arbiter PUF and RO PUF implementations used in the key generation and authentication scenarios. As a result, fault injection enables us to deactivate different PUF chains in an XOR Arbiter PUF, which simplifies a modeling attack against such architectures. In a similar way, we can disable arbitrarily chosen ring oscillators in different RO PUF variants to reduce and bias the entropy of the generated numbers.

5.1 Attack Scenarios

5.1.1 LFI Attack Against XOR Arbiter PUFs

Although different studies have revealed the vulnerability of XOR Arbiter PUFs with a limited number of Arbiter chains to ML attacks, an XOR Arbiter PUF with a large number of Arbiter chains is thought to be still secure against such attacks [70, 72]. Moreover, the complexity of current ML attacks on XOR Arbiter PUFs increases exponentially with the number of Arbiter chains. Here we elaborate briefly on an example of how the LFI can be combined with a well-established ML framework to break the security of an XOR Arbiter PUF. If the attacker could access the response of each individual Arbiter PUF in an XOR Arbiter PUF, each chain can be modeled separately in polynomial time, e.g., following the procedure in [26]. By obtaining a model of the challenge-response behavior of each Arbiter PUF individually, the challenge-response of an XOR Arbiter PUF can be predicted, and hence, the security of the XOR Arbiter PUF is broken. Two possible approaches can be deployed to

© Springer International Publishing AG, part of Springer Nature 2019

S. Tajik, *On the Physical Security of Physically Unclonable Functions*, T-Labs
Series in Telecommunication Services, https://doi.org/10.1007/978-3-319-75820-6_5

launch this attack. First, the attacker can inject a fault into the LUT, which realizes the XOR gate, and transform its configuration to a buffer gate for its first input, which is connected to the first Arbiter PUF. In this case, the response of the XOR Arbiter PUF is equal to the response of the first Arbiter PUF, and therefore, the attacker can conduct an ML attack and obtain the model of the first Arbiter PUF in a few seconds. To have the previous and unchanged circuit configuration again, the device has to be rebooted. After each reboot event, in a similar fashion, the XOR gate has to be reconfigured to a buffer for each and every input, which enables the attacker to model all individual PUFs. However, reconfiguration of an LUT to convert an XOR gate to a buffer requires a high precision laser shot and precise knowledge of the hardware architecture, which might not be feasible in a real scenario.

Another approach is to induce a fault into all the individual Arbiter PUFs to deactivate them, except the one which has to be learned. To this end, it is convenient to target the inverter chain connected to the clock input of Arbiter flip-flop of each single Arbiter PUF, see Fig. 5.1. By reconfiguration of an arbitrarily chosen inverter in this chain to another gate, which ignores the original input as *don't care*, the enable signal will not propagate through the chain anymore. If no signal reaches the clock input of the flip-flop, no sampling occurs, and the Arbiter PUF will respond with *zero* to all challenges. By applying this technique, each time after rebooting the device, the attacker can deactivate all Arbiter PUFs except one to learn it individually. Manipulating the configuration of an LUT to convert an inverter to a gate, which ignores the original input, can be carried out with a high probability without high precision laser shots and exact knowledge of hardware architectures.

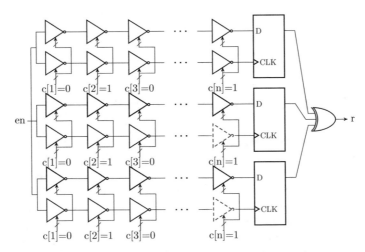

Fig. 5.1 The inverters marked with dashed lines are potential targets for an LFI attack against the XOR Arbiter PUF

It has been proven that an Arbiter PUF can be learned under a Deterministic Finite Automaton (DFA)-based representation by collecting only a polynomial number of CRPs for given accuracy and confidence levels (i.e., *PAC learning*) [26]. The accuracy level shows the error of the model whereas the confidence level represents the likelihood of delivering the model. When launching our above mentioned hybrid attack, each Arbiter PUF (e.g., ith PUF) can be PAC-learned individually for given accuracy and confidence levels, namely, ε_i and δ_i, respectively. According to the PAC model, ε_i can be set to a significantly small value (more precisely, close to zero) so that a virtually completely correct model would be delivered. Furthermore, δ_i can also be sufficiently small (i.e., close to zero), which means that with very high probability (i.e., close to one) the model would be delivered. Therefore, when *combining* the individually modeled Arbiter PUFs into a single model, the resulting total error of the joined model for the complete XOR Arbiter PUF (ε) will also be significantly small, as we are combining only a few modeled Arbiter PUFs. Moreover, the probability of delivering such a model ($1 - \delta$) will also still be close to one. In this work, we do not focus on the proof of the correctness of this setting and refer the reader to [40] for the proof and more details. Instead, we demonstrate how the LFI can assist us to reduce the number of CRPs required to model an XOR Arbiter PUF.

To compute the maximum number of CRPs required to PAC-learn an individual Arbiter PUF, we follow the procedure introduced in [26]. To establish the DFA-based representation, by applying a statistical discretization and a mapping process, the real-valued delays are mapped to integer-valued delays lying within $[0, M]$, where M is the maximum variation of delay values. The size of the DFA obtained regarding these processes is polynomial in the number of stages (n) and M. Therefore, the direct corollary of this and the theorem proved by Angluin [7] is that an Arbiter PUF (e.g., ith chain) can be PAC learned by collecting the maximum number of CRPs:

$$N_i = O\left(\left(1 + \frac{2}{\varepsilon_i}\ln(1/\delta_i)\right)nM^2 + \frac{2}{\varepsilon_i}n^2M^4\right).$$

Hence, if we assume that the accuracy and confidence levels are the same for all Arbiter chains (i.e., $\varepsilon_1 = \varepsilon_2 = \cdots = \varepsilon_k = \varepsilon$ and $\delta_1 = \delta_2 = \cdots = \delta_k = \delta$), the maximum total number of CRPs required for our hybrid attack is

$$N = \sum_{i=1}^{k} N_i$$

$$= O\left(\left(1 + \frac{2}{\varepsilon}\ln(1/\delta)\right)knM^2 + \frac{2k}{\varepsilon}n^2M^4\right)$$

where k is the number of Arbiter chains. By recalling the theorem proved in [26], we can summarize the above mentioned discussion in the following theorem.

Theorem 5.1 *To PAC learn an XOR Arbiter PUF with an arbitrary number of Arbiter chains k, when launching the hybrid attack, a polynomial-time algorithm can be found that requires at most N CRPs to return an approximated model of the PUF with a probability of at least* $(1 - \delta)$. *The number of CRPs N is polynomial in n, M as well as k.*

It is also stated that the same approach can be applied to PAC learn XOR Arbiter PUFs as target concepts [26]. To this end, the following maximum number of CRPs should be collected (for the detailed proof see [26]).

$$
N_{XOR} = O\left(\left(1 + \frac{2}{\varepsilon}\ln(1/\delta)\right)n^k M^{2k} + \frac{2}{\varepsilon}n^{2k}M^{4k}\right)
$$

From the last two equations, it can be concluded that by combining the LFI attack with the PAC learning attack, a significantly smaller number of CRPs needs to be collected by the adversary.

5.1.2   LFI Attack Against RO PUFs

RO PUFs are preferred in random number generation applications, due to the higher entropy density of their responses [48]. To negatively influence the entropy of the RO PUF response to different challenges, three combinatorial logic parts of the PUF can be attacked. One way is to manipulate the challenge multiplexer. In this case, the desired ring-oscillators cannot be selected by the challenges, and therefore, the PUF response will be affected. However, in a real implementation of an RO PUF with a large number of ring-oscillators, the challenge multiplexer is realized by multiple smaller multiplexers. Moreover, similar to the case of XOR gate reconfiguration discussed in the previous section, the exact reconfiguration of multiplexers requires precise knowledge of underlying architecture, and hence, it is hardly feasible.

Another possible way is to induce a fault into the configuration of one arbitrarily chosen inverter in the chain of the individual ring-oscillators, see Fig. 5.2. An induced faulty configuration will not invert the input signal with a high probability. Thus, the ring-oscillator stops oscillating and gives a constant value at its output. When there are two or more deactivated ring-oscillators, the binary response to a given challenge using these two ring-oscillators is generated as in the case of having two, or even more, equal frequencies in the PUF. This case must have been resolved by the manufacturer in the enrollment phase. Otherwise, the PUF response in the verification step will be undefined.

Comparing the frequencies between a pair of oscillators is the primary source of the entropy in the RO PUFs. There are $N!$ possible frequency orderings in an RO PUF with N ring-oscillators [79]. As a result, the entropy of such RO PUF will be $\log_2(N!)$ bit. The following theorem presents how the LFI attack can dramatically reduce the entropy of the RO PUF.

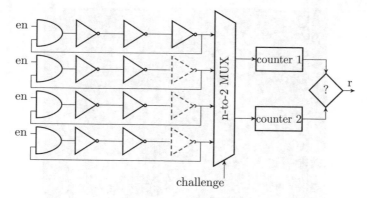

Fig. 5.2 The inverters marked with dashed lines are potential targets for LFI attack against the RO PUF

Theorem 5.2 *Assume that the attacker can randomly deactivate i ring-oscillators, out of N ring-oscillators implemented in the PUF. The entropy of the RO PUF consisting of the remaining operating ring-oscillators is* $\log_2((N-i)!)$ *bit.*

This can be easily proved due to the fact that the number of possible frequency orderings is $(N-i)!$ after launching the attack.

5.2 Results

In order to evaluate the feasibility of the LFI attack against PUFs, an RO PUF with 3 ring oscillators and an XOR Arbiter PUF with 2 Arbiter chains have been implemented on the Altera MAX V CPLD (see Sect. 3.1.1). The LFI attack would be successful, if the generated faulty configurations in the LUT block the signal propagation in a ring-oscillator or an inverter chain. Besides, the attack on PUFs would be effective, only if the targeted blocks are reconfigured without affecting other parts of the PUF or the chip functionality in general. Therefore, we have connected the output of all ring-oscillators and Arbiter PUFs directly to the output pins of the chip to monitor the behavior of all blocks at the same time. To find points of interest, we have captured photonic emission images with the CCD camera on the setup (see Sect. 3.3.2). As each hardware primitive has its emission fingerprint [85], we could identify and locate ring-oscillators and inverter chains on the chip, see Fig. 5.3. As soon as the location of the ring oscillators or inverter chains was determined in this way, the sensitive regions in the LABs and LEs of interest were then discovered by our scanning approach, see Sect. 3.3.2.

With these methods, faults are successfully induced into an arbitrarily chosen inverter of each ring-oscillator by manually triggering the laser, when the laser is scanning the sensitive regions of an inverter LUT. As a result, we were able to

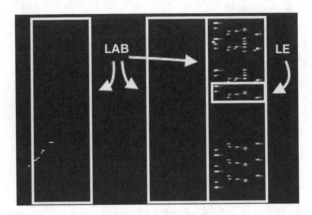

Fig. 5.3 Photonic emission image of a ring-oscillator acquired from CPLD backside

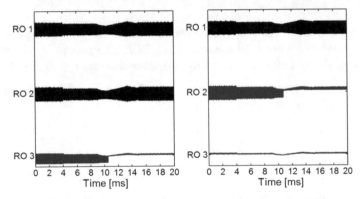

Fig. 5.4 The outputs of three ring-oscillators. LFI into an arbitrarily chosen inverter of the ring-oscillator 3 stops the oscillation of that ring, while the other two rings continue oscillating. In the second step, another fault was induced in one of the inverters of the ring-oscillator 2, which leads to deactivation of that ring. The behavior of the ring-oscillator 1 remained unchanged after a transient time

deactivate individual ring-oscillators by reconfiguring the inverter, while the other oscillators continued oscillating, see Fig. 5.4. After each successful fault injection, we could navigate the optical objective to the next ring-oscillator by reducing the laser power. In this case, although the laser had enough power to acquire an image from the surface of the chip for navigation, it was not powerful enough to induce undesired faults into the circuit. In the second step, we were able to deactivate the second ring-oscillator in a similar fashion, see Fig. 5.4.

The same approach can be applied to deactivate the clock inverter chain of an Arbiter PUFs. A fault can be induced to any arbitrarily chosen inverter in the chain. To monitor the behavior of the inverter chains, we have connected the enable signal of the PUFs to a clock source with 500 Hz. Hence, the same clock signal can be observed

Fig. 5.5 Outputs of two inverter chains, which are connected to the clock signal. By inducing a fault into an arbitrarily chosen inverter in the inverter chain 2, the signal path is blocked and the clock signal can never reach the end of the chain. The inverter chain 1 still functions properly after the deactivation of the other chain

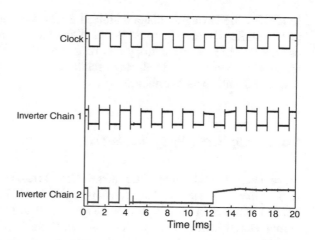

at the end of the chain with a few picoseconds delay, see Fig. 5.5. By inducing a fault into one of the inverters of the inverter chain, the chain was deactivated, and the output of the chains becomes a constant value.

5.3 Discussion and Potential Countermeasures

5.3.1 Scalability of the Attack

In Sect. 5.1, we provided a detailed theoretical analysis to evaluate the efficiency of the proposed hybrid attack (i.e., the combination of LFI attack and modeling attack). However, to gain more insight into that, we review a few relevant practical results reported in the literature. An Arbiter PUF with 128 stages can be modeled with 99% accuracy in 0.51 s using 5570 CRPs, while an 5-XOR Arbiter PUF with the same number of stages can be modeled with the same accuracy in more than 16 hours using 500000 CRPs [70]. A light weight secure PUF, which is an extended variant of the XOR Arbiter PUF, with 5 Arbiter PUFs can be modeled in 267 days using 10^6 CRPs [70]. In contrast to the pure ML attacks, the proposed hybrid attack in this chapter enables the attacker to model individual Arbiter PUFs independently in less than 1 s. Consequently, XOR or light weight secure Arbiter PUFs can be modeled in a few seconds with far less number of CRPs. It has been reported in the literature that the combination of ML attack and side-channel analysis can reduce the time and the number of CRPs required to model an XOR Arbiter PUFs with a limited number of Arbiter chains [72]. However, the proposed attack in this chapter can be applied to XOR Arbiter PUFs with an arbitrarily large number of Arbiter PUFs.

We provide an example for the second proposed attack against RO PUFs to demonstrate how it influences the entropy of PUF-based random number generators. If an

RO PUF with 512 ring-oscillator is utilized as a random number generator the entropy of the PUF response will be 3875 bit. If the attacker deactivate half of the ring-oscillators, the response entropy will be decreased to 1683 bit. In the worst case, the attacker can deactivate all ring-oscillators to get a constant and deterministic response for all applied challenges.

5.3.2   Applicability of the Attack

Since the LUT architecture of the Altera's CPLDs and FPGAs are virtually identical, in principle the same laser fault attack is applicable to FPGAs. In principle the proposed LFI attack in this chapter can be applied to all PUF instances, which contain combinatorial logic primitives. For instance, Bistable Ring (BR) PUFs are realized by a large number of NOR gates, multiplexers and demultiplexers [18]. Recently, the concept of an XOR BR PUF was introduced, which is believed to be secure against modeling attacks. However, the same LFI attack against XOR Arbiter PUFs, can be launched against individual BR PUF chains of such architecture to deactivate them one by one, and then learn them individually [93]. Finally, the behavior of other similar intrinsic hardware primitives, such as RO-based True Random Number Generators (TRNGs), can be affected by the proposed LFI attack. The main difference between an RO PUF and an RO-based TRNG is the lack of a challenge-response mechanism in the latter case. In an RO-based TRNG, all ring-oscillators are connected to an XOR gate. By sampling the output of the XOR gate in predefined intervals a random bit sequence can be obtained. To attack such TRNG, the attack introduced in Sect. 5.1.2 can be applied to arbitrarily chosen inverters in each of the ring-oscillators of the TRNG. As a result, the rings can be deactivated leading to decrease of the entropy at the output of TRNG.

5.3.3   Countermeasures

SRAM-based CPLDs and FPGAs can be protected from transient and permanent faults by applying different fault-tolerance techniques. Triple modular redundancy (TRM) and a combination of duplication with comparison (DWC) with concurrent error detection (CED) are examples of such techniques [39, 43]. In both cases, utilization of redundant LEs for the same operation and the majority voting on the outputs of them can lead to error-free outputs. However, duplication (i.e., physical cloning) of PUF components is hardly feasible on the chip. Therefore, although these countermeasures are able to avoid the negative effects of faults on the modular arithmetic operations, they are ineffective to protect the PUFs.

Error correction blocks, which are correcting the noisy responses of the PUF can be a potential countermeasure against the LFI attack against XOR Arbiter PUFs. Since deactivation of multiple chains of an XOR Arbiter PUF leads to alteration of the generated responses of the PUF, many responses are considered noisy for the error correction codes. If the nimber of noisy responses are more than a threshold, an alarm can be raised, and anti-tamper reaction can be carried out Moreover, the entropy of the generated responses of an RO PUF can be tested on the chip directly. Similarly, if the entropy test of the PUF responses fails, an alarm can be raised.

Another way is to modify the PUF construction itself to include additional logic components to monitor the healthiness of the PUF [73]. For instance, the outputs of the final stage of each Arbiter PUF in an XOR Arbiter PUF can be compared and checked to verify the arrival of the enable pulse. If an Arbiter PUF is deactivated by an LFI attack, the enable pulse cannot arrive at the end of the chain, and hence, the attack is detected. Furthermore, the reconfiguration capability of the modern reconfigurable hardware can be employed to reconfigure the correct PUF configuration into the chip upon detecting a reconfiguration attack [73]. Finally, integrity checking of the configuration memory during runtime raises the chances of detection of such attacks.

The lockdown protocol [95], which is briefly discussed in Sect. 4.3.4, can be deployed to protect the XOR Arbiter PUF against LFI attacks as well. Although this protocol cannot prevent the LFI attack directly, it can stop the attacker to obtain enough number of CRPs. Hence, the adversary cannot launch an ML attack against the PUF.

Chapter 6
Optical Contactless Probing

In this chapter, we demonstrate that all Intrinsic soft and hard PUF implementations in reconfigurable hardware, regardless of their architecture, are vulnerable to optical contactless probing. Since in a real scenario the implemented soft or hard PUFs inside of FPGAs are controlled PUFs, a non-invasive access to the CRPs of the PUFs is restricted by either physical or algorithmic countermeasures. Hence, most of the reported modeling techniques [10, 25, 71] and semi-invasive techniques [54, 64, 82, 84], including EM, PEM, and LFI, are ineffective to attack the PUF. In this fashion, the unprocessed challenges can be transferred within the FSBL to the FPGA, which is processed later on the device by non-linear functions and applied to the PUF. The response of the PUF will also be generated and processed inside the device and cannot be observed in a non-invasive way.

We show how the attacker can deploy LVI to locate circuitry of interest, such as key registers and ring-oscillators of an RO PUF, by knowing or estimating the frequency of different operations. We further present how LVP enables us to probe *volatile* and *on-die-only* data streams on the chip without having any physical contact to the transistors or wires. Furthermore, one can perform LVP to characterize high frequency signals, such as the output of ring-oscillators of an RO PUF. For our practical evaluation, we consider a PoC RO PUF implementation in key generation mode inside the FPGA.

We further propose an approach to using PUFs as physical sensors to monitor the integrity of reconfigurable hardware against LVP and LVI attacks. A few modifications in an existing PUF architecture enable us to design a PUF-based security scheme, which can be deployed for attack detection and authentication/key generation at the same time. We evaluate the effectiveness of our prototype scheme against optical contactless probing. Finally, we discuss how this scheme can be deployed during bitstream configuration in FPGAs with partial reconfiguration capability.

This chapter with slight revisions are based on publications [45, 83].

© Springer International Publishing AG, part of Springer Nature 2019
S. Tajik, *On the Physical Security of Physically Unclonable Functions*, T-Labs
Series in Telecommunication Services, https://doi.org/10.1007/978-3-319-75820-6_6

6.1 Attack Scenario

The principle of key generation inside an FPGA has been discussed in Sect. 2.2.1. The attacker can probe directly the red, black and PUF key using LVI and LVP. Extracting the red key enables the adversary to decrypt the encrypted bitstream offline, which make cloning of the design feasible. Moreover, if an RO PUF is deployed in the FPGA the attacker can characterize its ring-oscillators based on a combination of LVI, LVP, and power analysis. High precision frequency measurement of individual ring-oscillators enables the attacker to characterize the RO PUF. The main assumption of these attacks is a knowledge of the *approximate* location of the key registers and the PUF components on the FPGA.

6.1.1 Key Extraction

Based on the implementation all three key values (i.e., black key, PUF key, and red key) can be either loaded into the registers in parallel or loaded serially through a shift register, see Fig. 6.1. The attacker can use LVI directly to extract all three values if the keys are loaded and processed in parallel, Since LVI reveals nodes switching with a certain frequency or having certain frequency components (see Sect. 2.3.3), the adversary needs to take the switching frequencies of the red key registers during red key generation into account. If it is not predefined, all registers of an FPGA are first initialized to their default value (e.g., zero) by the reset circuitry after power-on. In this case, the black key registers are filled in parallel with the black key bits and the PUF circuit is activated. In a similar way, the PUF registers are filled in parallel and simultaneously with response bits of the PUF. In the final step, when the red key bits are available at the XOR outputs, they can be loaded into red key registers. As a result, we can observe that all register blocks (i.e., black key, PUF key, and red key) receive data exactly once per power-on. This fact can be used to generate desired frequency components by rebooting the device in a loop. Therefore, as the

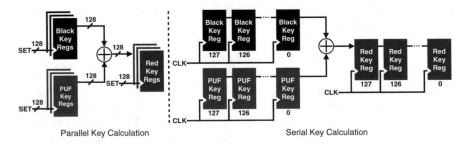

Fig. 6.1 a Parallel and serial generation of the red key

registers change their state once per reset, the first harmonic of the waveforms on these registers is the reboot frequency.

Figure 6.2 shows that there is a difference between the waveforms of two registers receiving a one and a zero bit. For the register receiving the bit "1" (i.e., REG_A) it is evident that the register starts at the logic level low and then changes its state to the logic level high, as soon as the time required for the former calculations (T_{CALC}) has elapsed. The register is reset when the reset input goes high, and afterward, the power-on cycle is restarted once reset goes low again. Since we T_{CALC} is constant for consecutive power-ons, the REG_A's period is T_{RST} and the first harmonic is at $1/T_{RST}$. For register carrying a "0" (i.e., REG_B) the case is straightforward. REG_B does not change its value at all, and hence, it has not any harmonics at the reset frequency. Therefore, the adversary can expect that the registers carrying a "1" to modulate the reflected light with a first harmonic of $1/T_{RST}$. Registers carrying a "0" do not modulate the reflected light at all. The interaction is similar for black key, PUF key and red key register blocks. Although T_{CALC} changes for different register blocks, the first harmonic for all of them is still at $1/T_{RST}$. Consequently, to read out the stored keys in a register block the attacker can perform LVI on the register block of interest, while setting the spectrum analyzer filter frequency to the reset loop frequency. If the LVI measurement is then grayscale encoded, registers carrying a "1" are expected to show up white while registers carrying a "0" will remain black.

The serial implementation creates a different situation. In this case, the data is processed bit by bit, and the individual registers in the relevant register blocks are connected together to form a shift register for each block. The stored black key and PUF key bits are then shifted out of the black key and PUF key shift registers, passed through the XOR and shifted into the red key shift register. Thus, each individual register would demonstrate a different waveform depending on its position in the shift register and the actual bit values. The waveforms of the individual registers would still have the reset frequency as their first harmonic. However, detecting the bit values can not be broken down to a simple black/white distinction as for the parallel case. Nevertheless, the attacker still detects and localize the registers of interest in an LVI image, although with varying signal strength. After successful localization of

Fig. 6.2 Waveforms of the reset signal (RST) and two registers, receiving a one (REG_A) and a zero (REG_B) bit

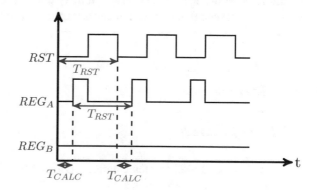

registers the attacker can move on to find the first register of each shift register, and probe the waveforms of individual registers directly using LVP. As the complete red key bits are shifted through the first register during calculation, the attacker can extract the red key from its waveform. Therefore, the attacker can extract the secret key regardless of the chosen implementation using only LVI or a combination of LVI and LVP.

6.1.2 RO PUF Characterization

To characterize an RO PUF, the frequencies of the ring oscillators has to be measured with high precision. Characterization of the RO PUF enables the attacker to clone its behavior. If the attacker can approximately estimate the frequency of the ring-oscillators, she can directly perform LVI measurement at that particular frequency. An approximation of frequencies can be obtained by EM or power analysis in the frequency domain. However, conducting SCA does not necessary reveal the frequency of individual ring-oscillators, but rather the superposition of all running ring-oscillators. Nevertheless, if she performs an LVI measurement at the approximate frequency with a large enough bandwidth, she should be able to observe the nodes of the ring-oscillators in the LVI image. As soon as the nodes of the ring-oscillators are identified and localized on the chip using LVI, the attacker can start to probe them individually. However, since the ring-oscillators are free-running, no trigger signal is available for waveform acquisition, and therefore, conventional LVP fails. As a solution, while probing one individual ring-oscillator, the attacker is free to connect the reflected light signal of the LVP directly to the spectrum analyzer of the LVP/LVI setup. Through setting the spectrum analyzer to conventional frequency sweep mode, she can then observe the spectrum of the reflected light signal. As the laser beam will just probe one node of one ring-oscillator, the waveform of the target ring-oscillator is modulated into the reflected light signal. Thus, the precise frequency of that individual ring-oscillator becomes visible on the spectrum analyzer. This technique eliminates the need for a trigger signal, and allow the attacker to characterize that specific ring-oscillator. By pointing the laser at the nodes of the remaining ring-oscillators, the attacker can then proceed to characterize the whole RO PUF.

6.2 Results

6.2.1 Key Extraction

We deployed a parallel implementation, as described in Sect. 6.1.1, for our first measurements. The black key was set to 10101101, the PUF key to 11011011 and the resulting red key was 01110110. The measurement was carried out with 5 MHz reset frequency and 50 MHz clock. Both signals had 50% duty cycle and 2.4 V high level and 0 V low level. The laser power was set to 10% and the pixel dwell time to 3.3 ms. The filter frequency for LVI was set to the reset frequency and the bandwidth to 300 Hz.

First, we performed an overview LVI image of an area containing all three register blocks, see Fig. 6.3a. There are clearly nodes whose waveforms contain frequency components at the reset frequency, and therefore, give rise to an LVI signal. Since it is known in which LABs of the FPGA the black key, PUF key and red key registers have been placed, it is now straightforward to assign the blocks to their respective keys. A higher resolution is helpful to analyze the data content of the registers. Hence, we repeated the LVI measurement on each register block while applying a scanner zoom. The resulting LVI images can be seen in Fig. 6.3b and the expected behavior discussed in Sect. 6.1.1 is observed. As expected, registers carrying "0" do not contribute to the LVI signal while registers carrying "1" can clearly contribute. We can observe slight differences in the appearance of the nodes in different measurements, which are probably due to a focus drift. Nevertheless, it is clear from the measurements that the attacker can easily extract the relevant values of the black key, PUF key and red key directly from these LVI images.

For the serial implementation, we utilized the same measurement setup. However, since the serial implementation requires more clock cycles to execute, the reset signal and LVI frequency were set to 1 MHz. The reset duty cycle was set to 58% as a makeshift trigger delay, causing only full bits to show up in the result before reset assertion. The laser power was increased to 15% and the pixel dwell time decreased to 1 ms. Afterward, an LVI image of the red key register block was taken, see Fig. 6.4. It is evident that there is no simple black/white data dependency, as discussed in Sect. 6.1.1. However, we can still observe a difference in signal strength for the

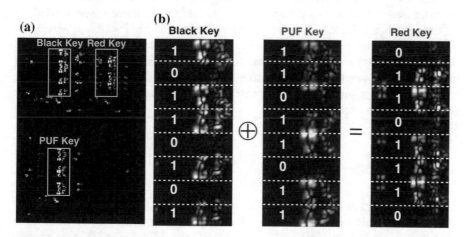

Fig. 6.3 LVI images of the parallel implementation. **a** All three register blocks taking part in the red key calculation. **b** Detail view of the individual register blocks. Dashed lines denote the LE boundaries. Each LE is approx. 6 μm in height

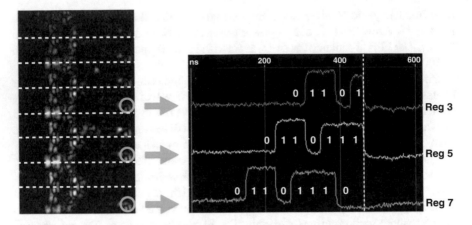

Fig. 6.4 LVI image of the red key register block and probed waveforms by LVP for the serial implementation. Reset assertion is marked by a dashed vertical line

registers, with the ones at the top of the Fig. 6.4 giving less signal than the ones at the bottom. To find out which points could be promising to perform LVP, we utilized fast Fourier transform (FFT) to analyze the amplitude of the first harmonic component for different expected waveforms. As a result, the waveforms with more bit shifts give us a stronger first harmonic component. Our conclusion was, therefore, that the lower half area of the LAB was the most promising location to probe. Direct probing of the lower-half registers was successful and revealed the lowest register to be the "shift-in" register. However, it was observed that waveforms with a better signal to noise ratio could be acquired at the locations close to the actual register area. We believe that these locations are associated with routing, and therefore, the signal has already been buffered before reaching them. Hence, the final measurements were carried out at these locations. The resulting waveforms can be seen in Fig. 6.4. It is apparent that the red key can be extracted from the lowest LVP waveform of the shift-in register by the attacker. We acquired further waveforms while setting the integration number down to 100000 loops, which is the current limit in the PHEMOS software, and were still able to distinguish the bit states easily. Therefore, we expect this approach to work with even fewer loop counts.

6.2.2 RO Characterization

For characterisation of the ring-oscillators, we applied the approach discussed in Sect 6.1.2. In this section, we present the frequency measurement for one of the ring-oscillators. We first deployed the Software Defined Radio (SDR) to get a rough estimation of the LVI frequency by measuring the superposition of all ring-oscillator's frequencies in the spectral domain on the power rail. By slight adjustments, we could

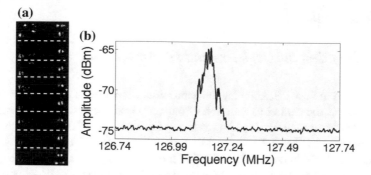

Fig. 6.5 **a** LVI image of 8 LEs of an RO, each approx. 6 μm in height. Dashed lines denote the LE boundaries. Each LE shows multiple potential probing locations. **b** LVP spectrum of the same RO

then create LVI overview images of the LEs forming the different ring-oscillators, one of which is depicted in Fig. 6.5a. For this LVI measurement, the spectrum analyzer filter frequency was set to 127.3539 MHz, laser power to 60%, and pixel dwell time to 0.33 ms. The ring-oscillators have much more short term frequency fluctuations than the previously used conventional clock sources. Therefore, the LVI filter bandwidth had to be set to 100 kHz to account for the more widespread ring-oscillator spectrum. After identifying the nodes of interest inside the LEs by applying LVI, the laser beam was held stationary on one of nodes, and the preamplified light detector signal was fed into the spectrum analyzer. Afterward, the spectrum analyzer was configured to show the spectrum of this signal, which was modulated by the ring-oscillator waveform present at the electrical node. For this measurement the laser power was set slightly higher, to 73%, the spectrum analyzer frequency span to 1 MHz, resolution bandwidth to 30 kHz and video bandwidth to 10 Hz. The resulting spectrum in Fig. 6.5b shows the ring-oscillator's frequency approximately 10 dBm above the noise floor. Thus, the attacker can determine the current ring-oscillator's frequency precisely using only contactless optical probing techniques.

It should be noted that the resolution bandwidth mentioned before is not the resolution to be expected for the frequency measurement. As the attacker will only be interested in the average frequency of the ring-oscillator, she is free to use multiple frequency sweeps to get a smooth spectrum and determine its peak value. The frequency of this peak value will then deliver the average frequency with a precision only depending on the number of averaged sweeps. By analyzing the average frequency acquired this way it can be seen that the ring-oscillator's frequency was shifted by approximately 0.15% when the laser power was increased from 60 to 73%. As long as the individual ROs are probed in the same way with the same laser power, this should not lead to problems for the attacker. Since the key issue for the attacker is just which ring-oscillator is faster, characterizing the RO PUF is still be successful if she takes care to probe all ring-oscillators in the same way, generating the same shift. Nevertheless, we will discuss this aspect in detail in Sect. 6.3.

6.3 Discussion

6.3.1 Locating the Registers and IP Cores on the Chip

As mentioned in Sect. 6.1, knowing the approximate location of the key registers and PUF IP core is the central assumption of our proposed attacks. Different scenarios can be considered to understand how realistic this assumption is.

As discussed in Sect. 2.2.1, the soft PUF IP cores, black key, and their placements are transmitted within the FSBL. If the FSBL or Boot0 is not encrypted, the attacker can intercept the boot loader on the board and gain knowledge about the configuration of the PUF and the red and black key registers. For instance, the Microsemi RoT solution [47] permits either the transfer of unencrypted or encrypted first stage boot loaders to the target SRAM-based FPGA. If the boot loader is encrypted, it will be decrypted by the hard dedicated AES core inside the target FPGA. While in the unencrypted case the boot loader can be easily intercepted, for the encrypted case DPA vulnerabilities of dedicated AES cores might be used to extract the encryption key and decrypt the boot loader [38, 60–62]. However, in the case of using asymmetric authentication and key rolling, as used by Xilinx SoCs, it is much harder for the attacker to expose the boot loader configuration [62]. Because of the authentication and key rolling, the attacker cannot launch a DPA attack against the hard AES core and therefore might not be able to decrypt the first stage boot loader.

If the first stage boot loader cannot be intercepted, the attacker has to have access to the used IP cores prior to the attack. Though difficult, it is imaginable that the adversary can get access to the IP cores via an insider or by posing as a potential customer to IP core suppliers. Having the IP cores, the attacker can synthesize the PUF on an identical FPGA model and analyze the design either in the IDE (if no obfuscation is used) or by looking at the generated bitstream to find the circuitry of the interest.

If the attacker cannot get access to the IP cores, the attack will be more difficult due to the unknown location of the circuitry of interest. In this case, if the utilized soft PUF is an RO PUF, one could launch the attack proposed in Sect. 6.1.2 to find the ROs and the counters connected to them on the chip. The location of the RO PUF can then be a reference point to localize other parts of the design inside the FPGA. Furthermore, one can estimate the operational frequency of different registers to apply LVI and localize the related registers individually on the chip. After a successful localization of the key registers, the attacker can extract data from them by LVP/LVI based on the implementation (See Sect. 6.2.1). In the case of a parallel implementation, if the key registers are naively implemented in the right order (i.e., from LSB to MSB), the attacker can easily extract the key by using LVI. Otherwise, if the keys are latched in an obfuscated way, the attacker can only read the state of the permuted registers and might not find the right order of the registers to assemble the key. For a serial implementation, if the order of the registers is obfuscated, the attacker can probe all registers to find the one through which the whole key is shifted.

The proposed attacks to key registers can in principle also be applied, if a hard PUF and a hard AES are in use. In this case, the attacker has to reverse-engineer the ASIC configuration circuit of the FPGA to locate the circuitry of interest. Although the search space for the region of interest might be reduced, the attacker has to probe and reverse-engineer more compact and dense ASIC circuits in comparison to FPGA logic cells, which might be challenging. Last but not least, it is obvious that LVP and LVI have the potential to directly probe the bitstream after on-chip decryption, circumventing all security measures in place.

6.3.2 Feasibility and Scalability of the Attack

The process technology of FPGAs and programmable SoCs, which are supporting partial reconfiguration for soft PUF implementation, are equal to or smaller than 60 nm. Since our LVI and LVP experiments have been carried out on an FPGA with 60 nm technology, the question of the applicability of the same technique on smaller technologies might be raised. The real size of the transistors is normally 7–8 times larger than the nominal technology node [31]. Besides, the size of the LEs and the routing (intra and inter LEs) of FPGAs is much larger than the size of the transistors, see Fig. 6.3. Hence, the optical resolution requirements for data extraction are much less severe than for probing individual transistors. Based on our measurements, the LE height in an Altera Cyclone IV is about 6 μm. The theoretical expected resolution of our laser spot is approximately 1 μm^2. Thus, optical probing should still be possible on an LE approx. six times smaller. It is worth mentioning that for LVP and LVI typical FPGAs are an advantageous target, as multiple transistors close together will carry the same waveform in an LE.

There are also solutions for increasing the optical resolution of LVP and LVI techniques. For instance, one can use solid immersion lenses (SILs) to get 2–3 times better resolution, which already enables *single transistor* probing at 14 nm [31]. Moreover, lasers with shorter wavelengths (e.g., in the visible light spectrum) can be used to further increase the resolution [12, 16]. However, in the latter case, the substrate of the chip has to be thinned to 10 μm or less to prevent the absorption of the photons.

Meanwhile, it is still interesting to understand why other backside semi-invasive attacks, such as PEM or LFI, have limited efficiency on small technologies in comparison to LVP and LVI. In the case of PEM, the photon emission rate is proportional to the core voltage of the chip. However, the core voltage of technologies smaller than 60 nm is too low [81] and the attacker therefore has to integrate over a large number of iterations to capture enough photons for analysis. LFI attacks on the other hand target mostly single memory cells, which requires the system used for the attack to be able to resolve single transistors on the chip.

6.3.3   Tamper Evidence

Although passive semi-invasive attacks do not affect the behavior of the PUF, the laser beam in our proposed attack can change the temperature of the transistors. Temperature variations have transient and reversible effects on the delay and frequency of the inverter chains in arbiter PUFs and RO PUFs. In our experiments, a shift of frequency has been observed while performing LVI and LVP on the ROs. However, the attacker is still able to precisely characterize and measure the frequencies of the ROs by performing LVI and LVP, if she takes care to probe all ring oscillators under the same conditions. If the attacker is not able to fulfill this requirement, she might also probe the registers of the counters which are connected to the RO output. Assuming the counters or other circuitry connected to the RO PUFs are located far enough away she will be able to mount her attack without influencing the ROs. Finally, she might take measurements of one individual RO frequency for different laser powers and extrapolate from that to the frequency for zero laser power. Therefore, a precise physical characterization of the RO PUF is certainly feasible.

6.4   Potential Countermeasure

Silicon light sensors have been proposed to detect the photons of the laser beam. However, in our experiments we have used a laser beam which has a longer wavelength than the silicon band gap. Hence, no electron-hole pairs will be generated by the laser photons. A silicon photo sensor is therefore unlikely to trigger.

A potential algorithmic countermeasure can be randomization of the reset states of the registers for the parallel implementation. As a result, the simple black/white data distinction (see Sect. 6.1.1) would be severely impeded, as there now would be switching activity during the reset loop on all registers. For the serial case, a randomization of the relation of the outer reset signal to the internal reset signal would destroy the needed trigger relationship and make waveform probing on the registers impossible. Another simple countermeasure includes the obfuscation of the key registers by randomizing their order, see Sect. 6.3.1.

Optical probing could have an immediate disturbance on temperature and current of the chip. The local temperature and current variations can affect the propagation delays of the electrical signals in the delay-based PUFs. Since soft PUFs have been already considered for key generation purposes inside the FPGAs [67], it is tempting to use the same PUFs as physical sensors to detect optical probing attacks as well.

6.4.1 Requirements for PUFs as Sensors

Optical probing could induce temperature and current variations into the chip. These local variations change the signal propagation delays of the transistors. Naturally, an ideal sensor should have a high resolution to measure tiny variations in the different physical parameters. However, to detect optical probing attacks by a sensor, other conditions have to be fulfilled as well.

Large spatial coverage Performing optical probing over a certain area of the chip only locally increases the temperature of the wires and transistors. Therefore, an ideal sensor should cover the whole area of the chip to detect local variations.

Large temporal coverage The irradiating of individual gates or registers during an optical probing attempt is potentially a very fast process. Hence, the sensor should continuously monitor the physical conditions of the chip to successfully detect an attack.

Security The attacker might make an effort to deactivate the sensor or tamper with the sensor output to hide her attack attempts. In an ideal case, any physical modification of the sensor should lead to an irreversible damage of the sensor.

6.4.2 Sensor Candidates

The behavior of timing-based circuits, e.g., delay-based PUFs [41] and ring-oscillator networks (RONs) [97], heavily rely on the propagation delays of their composing combinatorial components. To have a large spatial coverage, the ring-oscillators in a RON with virtually equal frequencies can be distributed all over the FPGA, see Fig. 6.6a. Therefore, performing local probing attempts will slightly shift the frequencies of ring-oscillators, which are in, or close to the probed area. The frequency deviation of affected ring-oscillators can be compared to the mean frequency of all ring-oscillators to detect the attack. Similarly, the individual ring-oscillators of RO PUFs can be realized in a distributed way on the chip.

Other delay-based PUFs (e.g., arbiter PUFs) can be employed inside an FPGA in a similar manner to cover the whole area of the chip. For instance, the multiplexers or the inverters of a large arbiter PUF can be placed and distributed manually all over the FPGA fabric. However, satisfying the temporal coverage requirements, especially for PUF-based sensors, is more challenging. Although the laser irradiation can instantly alter the delays of one or more PUF components, the affected components might not be active during the attack period. If we assume that an arbiter PUF is used as a sensor, by enabling the PUF for a specific challenge, each stage of the arbiter PUF is active only for a few picoseconds to pass the incoming signal from the previous stage to the next one. In this case, instant delay alteration of one stage prior or after the signal handover will not have any effect on the behavior of the PUF. Hence, a sensor with constantly active elements should be selected if a high temporal coverage is required.

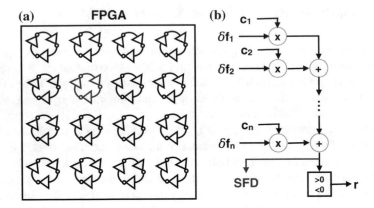

Fig. 6.6 **a** A distributed placement of the ROs inside the FPGA. Performing optical probing can shift slightly the frequency of an RO, which is in, or close to the probed area. **b** The modified architecture of the RO sum PUF. The summation of frequency differences is measured directly before making a decision about the output of the PUF

Among digital intrinsic hardware primitives, RO sum PUFs [94] and RONs can offer better temporal coverage since their ring-oscillators can be made always active to sense the anomalies.

A permanent physical modification of both RONs and delay-based PUFs by fully-invasive attacks with a high probability leads to the destruction of them. Moreover, their deactivation leads to altered outputs. Thus, they can be considered tamper-evident against fully-invasive attacks and satisfy the security condition.

6.4.3   Combining RO Sum PUFs and RONs

RONs have been used to create signatures, detect small current variations as well as hardware Trojans in ASICs. Upon activation of the ring-oscillators, the counters count the number of rising edges at the outputs of the ROs within a predefined interval. The frequency of each ring-oscillator is measured separately by a counter and summed up to be analyzed externally. Mounting active attacks can *directly* alter the frequency values, and the attack can be detected with a high probability. However, RONs are neither common IPs for FPGAs nor well suited for key generation and authentication simultaneously. The latter is due to the lack of both exponential input space and binary responses.

On the other hand, RO sum PUFs consist of n pairs of ring-oscillators, whose outputs are connected to binary counters. Similar to RONs, after a predefined period, the states of the counters of two adjacent ROs are sampled. The sampled values are then subtracted from each other to generate n frequency difference values. Based on an applied n-bit challenge to the PUF, each frequency difference value is multiplied by $+1$ or -1. Finally, all the values are summed, and the final response of the PUF is 0 if the sum is negative. Otherwise, it equals 1. An RO sum PUF has an exponential CRP space, and therefore, can be utilized for authentication and key generation purposes. However, it should be noted that changes in the frequencies of a small number of ring-oscillators that are under an active attack do not necessarily change the *binary* response of the PUF, which is undesired in the case of a PUF acting as an attack detection sensor.

To have an exponential CRP space and immediate changes in the response during active attacks, the concepts of RON and RO sum PUF can be combined. If the summation of the frequency differences (SFD) in the RO sum PUF is measured before the decision making (similar to a RON), instant changes in the behavior of ROs can be observed, see Fig. 6.6b. Meanwhile, if the binary responses of the PUF are not affected, the PUF can be utilized for key generation and authentication purposes. We refer to this PUF-based security monitoring scheme as PUFMon.

6.4.4 Enrollment and Verification

Similar to PUF-based authentication, a PUF-based physical integrity sensor has to be evaluated in two phases. First, in the enrollment phase for each PUF, a set of CRPs is measured and stored in a database in a trusted field. Later in the verification phase, the enrolled challenges are retrieved from the database and applied to the PUF, and the outputs generated by the PUF are compared to the enrolled outputs. Since the SFDs are used later in the field to detect an attack, the SFDs of the PUFMon are measured several times in a normal and fault-free condition to obtain their maximum and minimum values. These values can be gathered in parallel with the actual binary responses of the RO sum PUF during the enrollment phase in a trusted field. Afterward, in an adversarial field, an alarm can be raised if a predefined percentage of the SFDs does not lie within the min/max interval determined in this way.

There are different options for storing the SFD min/max limits of the PUF. One option is to transmit a set of enrolled limits encapsulated in the encrypted first stage boot loader (FSBL) or boot0 to the FPGA and store them in the volatile memories. This technique can only be applied to FPGAs with partial reconfiguration capability. In this scenario, the current SFDs of the PUFMon can be compared offline with the SFD limits stored inside the chip to detect anomaly conditions. Although offline verification can be effective against non-invasive active attacks during runtime, the SFD limits might be vulnerable to tampering by invasive attacks. Another option is to store the SFD limits externally in a RoT or a secure database. In this case, the FPGA should communicate with the database online and verify the behavior of the FPGA. Finally, in a hybrid approach, while the PUFMon monitors the integrity of the FPGA offline, a server can occasionally update the stored challenges and SFD limits on the chip to raise the security level. It should be noted that in an ideal case the communication between FPGA and the database should be secured by encryption and authentication schemes. Otherwise, the attacker might be able to characterize the PUF by intercepting the SFDs.

6.4.5 Monitoring of Optical Probing Attempts

We conducted our optical probing detection experiment with two different wavelengths. During the monitoring phase, we performed optical contactless probing over an area of the chip. Each step of our experiments consists of 10 rounds of SFD evaluation by applying the set of 100 enrolled challenges. We started our first 10 rounds of monitoring rounds without performing probing. Afterward, step by step we increased the power of the laser by 10% for 10 rounds. After each step, the laser was turned off for 10 rounds, see Fig. 6.7.

To interpret the experimental results, the effect of photons with different wavelengths on the transistors should be understood. The silicon substrate is more absorptive at 1064 nm than at 1300 nm, and thus, less 1064 nm photons reach the transistors. However, since the photons with the wavelength of 1300 nm contain less energy than the band gap of the silicon, they mainly have thermal effects. On the other hand, the photons with a wavelength of 1064 nm have higher energy than the band gap of the silicon, and hence, in addition to thermal effects they generate electron-hole pairs, which leads to current induction in the transistors. Stimulation of transistors with 1064 and 1300 nm photons is called Photoelectric Laser Stimulation (PLS) and Thermal Laser Stimulation (TLS), respectively [17]. As a result, the shift in frequency of the ROs is higher when using the 1064 nm laser, and the detection is more probable. As can be seen in Fig. 6.7, given enough laser power, there is detectable laser influence on SFDs. The maximum power for our 1064 nm laser experiment was 50%, because there was a risk of damaging the transistors permanently.

A phenomenon that can be observed during high power experiments with the 1300 nm laser is the shift of the SFDs even after the laser is powered off. Naturally, increasing the laser power leads to an increase in the amount of the heat deposited in the chip during the *laser-on* period. Therefore, a few SFDs of the RO sum PUF still behaves out-of-bound at the end of the *laser-off* period.

6.4.6 Strengths and Weaknesses of PUFMon

PUFMon can be employed in each and every generation of reconfigurable hardware, particularly on less expensive ones without security features commonly used in the field. Moreover, PUFMon can be implemented independent of the user application. Finally, the behavior of the PUFMon is entangled with a particular device. If the SFD limits of one platform are divulged, the PUFMon behavior of other devices remains unknown.

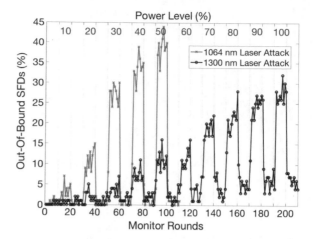

Fig. 6.7 The effects of the performing LVP/LVI with the 1064 and 1300 nm laser on the SFDs

To enable a successful key generation and authentication, the challenge-response behavior of the PUF has to be stable. We observed that despite large changes in the SFD values of the PUF under attack, most binary responses of the PUF remained intact. Accordingly, if our proposed modified RO sum PUF is considered as a soft PUF IP, it can be used for authentication, key generation and monitoring at the same time. Consequently, the PUF configuration can be transmitted from an RoT within the FSBL to FPGAs with partial reconfiguration capability. Later, the bitstream can be transferred from the RoT to the FPGA via partial reconfiguration. During partial reconfiguration, PUFMon can communicate with the RoT to verify the correct functionality of the FPGA.

Based on our observations, the SFDs are highly sensitive to small variations in the power supply voltage and global temperature, and thus, it can raise false alarms. For instance, by increasing or decreasing the core voltage of the FPGA by 10 mV, virtually all SFDs behave out-of-bound. As a solution, one can set the detection threshold higher at the cost of a lower detection probability for more local attacks. Another weakness of the PUFMon might be the high power consumption of its ring-oscillators, which could make it unattractive for low power applications. One could decrease the number of the ring-oscillators and place them only close to the critical IPs. However, the spatial coverage of the PUFMon will be reduced. Another solution is to activate the PUFMon only within the critical periods, e.g., during configuration and encryption/decryption phases. Besides, PUFMon could have standby times to decrease the overall power consumption. Nonetheless, in both cases, the temporal coverage of the PUFMon is reduced.

Chapter 7
Conclusion and Future Work

Reconfigurability, flexibility and lower time-to-market have made the reconfigurable hardware the platform of choice for designing embedded devices. To protect the running IPs on these devices from cloning and manipulations, several protection mechanisms have been integrated into these platforms to neutralize several categories of physical attacks. Undeniably, one of the fundamental security challenges for vendors is the providing a protection scheme against advanced semi- and fully-invasive attacks from the IC backside. While fully-invasive techniques, such as FIB microprobing [36], are taken seriously by the chip manufactures, not enough attention has been paid to optical semi-invasive attacks in the past.

Although FIB machines are more expensive than optical setups used in this work, they are available in many research laboratories around the world. This is due to their wide use not only in semiconductor industry but also in physics and material sciences, and therefore, they can be rented by an adversary with very low prices. By adding more obfuscation and redundant security fuses, vendors can reduce the probability of success of an adversary, who is equipped with a FIB machine. Moreover, vendors can integrate charge sensors to their chips to detect FIB attacks [33].

On the contrary, optical semi-invasive techniques may not widely be accessible, and hence, they are not considered threatening by the chip manufacturers. Additionally, the common belief is that with the help of optical attacks an adversary cannot achieve enough resolution to probe a signal from a transistor or actively influence single transistors on the chip to launch an attack. While it is true that scaling down a set of optical analysis techniques (e.g., PEM and LFI) to the very latest nanoscale technologies are an onerous task, it is still feasible to utilize them [76, 78]. Besides, it should be kept in mind, while the size of the transistors is shrinking, modern inexpensive failure analysis approaches are developed to debug and probe nanoscale manufactured circuits in a semi-invasive and contactless way. Optical contactless probing, including LVP and LVI techniques, is an example of such advanced methods. It is worth mentioning that much less time is required for optical contactless probing of different signals than for FIB microprobing. The amount of time needed to probe multiple nodes optically is on the order of minutes while for FIB microprobing

© Springer International Publishing AG, part of Springer Nature 2019   71
S. Tajik, *On the Physical Security of Physically Unclonable Functions*, T-Labs
Series in Telecommunication Services, https://doi.org/10.1007/978-3-319-75820-6_7

it will be on the order of days. Hence, the higher renting cost of optical equipment in comparison to FIB machines is compensated by the less time, which is needed to conduct an attack.

Another reason, which makes vendors less concerned about semi-invasive attacks from the IC backside is the sample preparation of chips from the backside, which is considered a challenging task. However, modern high-performance FPGAs and SoCs are offered in flip-chip ball grid array (BGA) packages. In contrast to traditional packaging, the silicon die is inverted and placed face down in the package in flip-chip BGA packaging. Hence, access of adversary to the backside of the chip have become less complicated in the case of flip-chips.

Naturally, vendors can deploy digital Intrinsic PUFs to raise the security of key storage against invasive attacks on the modern reconfigurable hardware, instead of modifying the package and embedding an analog sensor to detect the attack attempts. The latter solution is more expensive and more challenging to be integrated into these devices, which makes them the unpopular choice for vendors. Therefore, an NVM might be replaced with a PUF to make the secrets volatile, harder to extract and even tamper-evident to invasive attacks. Indeed, one can deploy delay-based PUFs and distribute its elements on the whole area of a chip to develop a sensor to detect the FIB microprobing attempts by observing the intrinsic behavior of the PUF [14, 74].

However, PUFs might be only tamper-evident to fully-invasive techniques and not semi-invasive ones. Based on the already published results in the literature [14, 34, 35, 64, 74] and results of this work, it has become apparent that digital Intrinsic PUFs are not sensitive to sample preparation and optical semi-invasive attacks in their current configurations. By launching different classes of semi-invasive attacks, we demonstrated that Intrinsic PUFs are vulnerable to these attacks their responses can be predicted, manipulated and probed. Hence, Intrinsic PUFs cannot be considered as an ultimate solution to protect chips from all classes of physical attacks, and more effective analog sensors have to be added to the ICs. We firmly believe that the future generations of reconfigurable hardware remain vulnerable to semi-invasive attacks if no proper protections or countermeasures for the IC backside are implemented.

Future Work. In this thesis, we have assessed the physical security of the PUFs by launching optical semi-invasive attacks against *PoC* PUF implementations. To obtain a better understanding of the threat in a real scenario, one should conduct the same experiments against real soft and hard PUF implementations on the FPGAs, where little or no prior knowledge of the design is available. This requires to having access to such devices and PUF implementations, which were not available to the author of this work at the time experiments. In such scenarios, multiple steps have to be taken to localize and reverse-engineer the circuits of interest.

We demonstrated that optical contactless probing is a powerful technique to probe the responses of a PUF. In principle, the same method can be deployed to directly probe the entire bitstream of the FPGA after authentication and decryption on the chip. Possessing the bitstream enables the attacker to clone the design and implement it on other FPGA platforms. The main challenge would be the localization of the

ASIC AES decryption core and its outputs on the FPGA with the help of LVI. After finding the points of interest, the attacker can probe the bitstream with the aid of LVP.

Based on the results of conducted experiments in this work, it becomes apparent that PUFs alone cannot protect the chip from semi-invasive attacks from IC backside. Hence, novel protection mechanisms for the IC backside should be researched. One potential physical protection could be the utilization of different coating layers on the silicon substrate, which can reflect the incoming photons from a set of light sources to set of light detectors [15]. In this structure, to unblock the optical path from transistors, the coating layers must be removed from the silicon substrate. As a result, the incoming photons from the light sources are not reflected anymore, and consequently, not received by the light detectors in the absence of the coating layers. Therefore, the attack attempts from the IC backside can be detected. The idea can be further developed to build an optical PUF on the backside of the chip, where the selection of light sources and the detection of photons from different angles by light detectors can be considered as challenges and responses, respectively. In this case, any modifications to the silicon substrate alter the challenge-response behavior of the PUF, which makes the detection of the attack feasible.

References

1. Amazon Web Services, Inc. https://aws.amazon.com/ec2/instance-types/f1/ (2015). Accessed 02 May 2017
2. Helion Technology Limited. http://www.heliontech.com (2017). Accessed 02 May 2017
3. Intrisic-ID Inc. https://www.intrinsic-id.com (2017). Accessed 02 May 2017
4. Verayo Inc. http://www.verayo.com (2017). Accessed 02 May 2017
5. Altera: MAX V Device Handbook. Altera Corporation, San Jose (2011)
6. Altera: Cyclone IV Device Handbook. Altera Corporation, San Jose (2014)
7. Angluin, D.: Learning regular sets from queries and counterexamples. Inf. Comput. **75**(2), 87–106 (1987)
8. Armknecht, F., Maes, R., Sadeghi, A., Standaert, O.X., Wachsmann, C.: A formalization of the security features of physical functions. In: IEEE Symposium on Security and Privacy (SP), pp. 397–412. IEEE (2011)
9. Bar-El, H., Choukri, H., Naccache, D., Tunstall, M., Whelan, C.: The sorcerer's apprentice guide to fault attacks. Proc. IEEE **94**(2), 370–382 (2006)
10. Becker, G.T.: The gap between promise and reality: on the insecurity of XOR arbiter PUFs. In: Cryptographic Hardware and Embedded Systems–CHES 2015, pp. 535–555. Springer (2015)
11. Becker, G.T., Kumar, R.: Active and passive side-channel attacks on delay based PUF designs. IACR Cryptology ePrint Archive 2014, 287 (2014)
12. Beutler, J.: Visible light LVP on bulk silicon devices. In: 41st International Symposium for Testing and Failure Analysis (November 1–5, 2015). ASM (2015)
13. Boit, C.: Fundamentals of photon emission (PEM) in silicon—electroluminescence for analysis of electronic circuit and device functionality. In: Microelectronics Failure Analysis: Desk Reference. p. 356 ff. ASM International (2004)
14. Boit, C., Kerst, U., Schlangen, R., Kabakow, A., Le Roy, E., Lundquista, T., Pauthnerb, S.: Impact of back side circuit edit on active device performance in bulk silicon ICs. In: International Test Conference, vol. 2, p. 1236 (2005)
15. Boit, C., Tajik, S., Scholz, P., Amini, E., Beyreuther, A., Lohrke, H., Seifert, J.: From IC debug to hardware security risk: the power of backside access and optical interaction. In: IEEE 23rd International Symposium on the Physical and Failure Analysis of Integrated Circuits (IPFA), pp. 365–369. IEEE (2016)
16. Boit, C., Lohrke, H., Scholz, P., Beyreuther, A., Kerst, U., Iwaki, Y.: Contactless visible light probing for nanoscale ICs through 10 μm bulk silicon. In: Proceedings of the 35th Annual NANO Testing Symposium—NANOTS 2015, pp. 215–221 (2015)
17. Brahma, S.K., Heinig, J., Glowacki, A., Leihkauf, R., Boit, C.: Distinction of photo-electric and thermal effects in a MOSFET by 1064 nm laser stimulation. In: 13th International Symposium on the Physical and Failure Analysis of Integrated Circuits, pp. 333–339. IEEE (2006)

18. Chen, Q., Csaba, G., Lugli, P., Schlichtmann, U., Ruhrmair, U.: The bistable ring PUF: a new architecture for strong physical unclonable functions. In: IEEE International Symposium on Hardware-Oriented Security and Trust (HOST), pp. 134–141. IEEE (2011)

19. Davidson, A.: WP-01220-1.1: A New FPGA Architecture and Leading-Edge FinFET Process Technology Promise to Meet Next-Generation System Requirements. Altera Corporation, San Jose (2015)

20. Delvaux, J., Gu, D., Schellekens, D., Verbauwhede, I.: Secure lightweight entity authentication with strong PUFs: mission impossible? In: Cryptographic Hardware and Embedded Systems–CHES 2014, pp. 451–475. Springer (2014)

21. Delvaux, J., Verbauwhede, I.: Side channel modeling attacks on 65nm arbiter PUFs exploiting CMOS device noise. In: IEEE International Symposium on Hardware-Oriented Security and Trust (HOST), pp. 137–142. IEEE (2013)

22. Delvaux, J., Verbauwhede, I.: Fault injection modeling attacks on 65 nm arbiter and RO sum PUFs via environmental changes. IEEE Trans. Circuits Syst I: Regul. Pap. **61**(6), 1701–1713 (2014)

23. Ganji, F., Krämer, J., Seifert, J.P., Tajik, S.: Lattice basis reduction attack against physically unclonable functions. In: Proceedings of the 22nd ACM SIGSAC Conference on Computer and Communications Security, pp. 1070–1080. ACM (2015)

24. Ganji, F., Tajik, S., Seifert, J.P.: Let me prove it to you: RO PUFs are provably learnable. In: Information Security and Cryptology-ICISC 2015. Springer (2015)

25. Ganji, F., Tajik, S., Seifert, J.P.: Why attackers win: on the learnability of XOR arbiter PUFs. In: Trust and Trustworthy Computing, pp. 22–39. Springer International Publishing (2015)

26. Ganji, F., Tajik, S., Seifert, J.P.: PAC learning of arbiter PUFs. J. Cryptogr. Eng. **6**(3), 249–258 (2016)

27. Gassend, B., Clarke, D., Van Dijk, M., Devadas, S.: Controlled physical random functions. In: Proceedings 18th Annual Computer Security Applications Conference, pp. 149–160. IEEE (2002)

28. Gassend, B., Clarke, D., Van Dijk, M., Devadas, S.: Silicon Physical Random Functions. In: Proceedings of the 9th ACM Conference on Computer and Communications Security, pp. 148–160. ACM (2002)

29. Guajardo, J., Kumar, S.S., Schrijen, G.J., Tuyls, P.: FPGA intrinsic PUFs and their use for IP protection. In: Cryptographic Hardware and Embedded Systems–CHES 2007, pp. 63–80. Springer (2007)

30. Güneysu, T., Markov, I., Weimerskirch, A.: Securely sealing Multi-FPGA systems. In: Reconfigurable Computing: Architectures, Tools and Applications, pp. 276–289. Springer (2012)

31. von Haartman, M.: Optical fault isolation and nanoprobing techniques for the 10 nm technology node and beyond. In: 41st International Symposium for Testing and Failure Analysis (November 1–5, 2015). ASM (2015)

32. Hansen, L.: White Paper WP470: Unleash the Unparalleled Power and Flexibility of Zynq UltraScale+ MPSoCs. Xilinx, Inc, San Jose, CA (2015)

33. Helfmeier, C., Boit, C., Kerst, U.: On charge sensors for FIB attack detection. In: IEEE International Symposium on Hardware-Oriented Security and Trust (HOST), pp. 128–133. IEEE (2012)

34. Helfmeier, C., Boit, C., Nedospasov, D., Seifert, J.P.: Cloning physically unclonable functions. In: IEEE International Symposium on Hardware-Oriented Security and Trust (HOST), pp. 1–6. IEEE (2013)

35. Helfmeier, C., Boit, C., Nedospasov, D., Tajik, S., Seifert, J.P.: Physical vulnerabilities of physically unclonable functions. In: Proceedings of the Conference on Design, Automation & Test in Europe, p. 350. European Design and Automation Association (2014)

36. Helfmeier, C., Nedospasov, D., Tarnovsky, C., Krissler, J.S., Boit, C., Seifert, J.P.: Breaking and entering through the silicon. In: Proceedings of the 2013 ACM SIGSAC Conference on Computer & Communications Security, pp. 733–744. ACM (2013)

37. Herder, C., Ren, L., van Dijk, M., Yu, M.D.M., Devadas, S.: Trapdoor computational fuzzy extractors and stateless cryptographically-secure physical unclonable functions. IEEE Trans. Dependable Secur. Comput. **2016**(99), 1–1 (2016)

38. Hori, Y., Katashita, T., Sasaki, A., Satoh, A.: Electromagnetic side-channel attack against 28-nm FPGA device. Pre-proceedings of WISA (2012)
39. Kastensmidt, F.L., Sterpone, L., Carro, L., Reorda, M.S.: On the optimal design of triple modular redundancy logic for SRAM-based FPGAs. In: Proceedings of the Conference on Design, Automation and Test in Europe, vol. 2, pp. 1290–1295. IEEE Computer Society (2005)
40. Kearns, M.J., Vazirani, U.V.: An Introduction to Computational Learning Theory. MIT press (1994)
41. Lee, J.W., Lim, D., Gassend, B., Suh, G.E., Van Dijk, M., Devadas, S.: A technique to build a secret key in integrated circuits for identification and authentication applications. In: Symposium on VLSI Circuits, Digest of Technical Papers, pp. 176–179. IEEE (2004)
42. Lim, D., Lee, J.W., Gassend, B., Suh, G.E., Van Dijk, M., Devadas, S.: Extracting secret keys from integrated circuits. IEEE Trans. Very Large Scale Integr. (VLSI) Syst. **13**(10), 1200–1205 (2005)
43. de Lima Kastensmidt, F.G., Neuberger, G., Hentschke, R.F., Carro, L., Reis, R.: Designing fault-tolerant techniques for SRAM-based FPGAs. IEEE Design Test Comput. **6**, 552–562 (2004)
44. Lohrke, H., Scholz, P., Boit, C., Tajik, S., Seifert, J.P.: Automated detection of fault sensitive locations for reconfiguration attacks on programmable logic. In: 42nd International Symposium for Testing and Failure Analysis (November 6–10, 2016). ASM (2016)
45. Lohrke, H., Tajik, S., Boit, C., Seifert, J.P.: No place to hide: contactless probing of secret data on FPGAs. In: Cryptographic Hardware and Embedded Systems–CHES 2016, pp. 147–168. Springer (2016)
46. Lu, T., Kenny, R., Atsatt, S.: White Paper WP-01252-1.0: Stratix 10 Secure Device Manager Provides Best-in-Class FPGA and SoC Security. Altera Corporation, San Jose, CA (2015)
47. Luis, W., Richard Newell, G., Alexander, K.: Differential power analysis countermeasures for the configuration of SRAM FPGAs. In: Military Communications Conference, MILCOM 2015-2015 IEEE, pp. 1276–1283. IEEE (2015)
48. Maes, R.: Physically Unclonable Functions: Constructions. Properties and Applications. Springer, Berlin (2013)
49. Maes, R., van der Leest, V., van der Sluis, E., Willems, F.: Secure key generation from biased PUFs. In: Cryptographic Hardware and Embedded Systems–CHES 2015, pp. 517–534. Springer (2015)
50. Maes, R., Verbauwhede, I.: Physically unclonable functions: a study on the state of the art and future research directions. In: Towards Hardware-Intrinsic Security, pp. 3–37. Springer (2010)
51. Majzoobi, M., Koushanfar, F., Devadas, S.: FPGA PUF using programmable delay lines. In: IEEE International Workshop on Information Forensics and Security (WIFS), pp. 1–6. IEEE (2010)
52. Mangard, S., Oswald, E., Popp, T.: Power Analysis Attacks: Revealing the Secrets of Smart Cards (2008)
53. Merli, D., Heyszl, J., Heinz, B., Schuster, D., Stumpf, F., Sigl, G.: Localized electromagnetic analysis of RO PUFs. In: IEEE International Symposium on Hardware-Oriented Security and Trust (HOST), pp. 19–24. IEEE (2013)
54. Merli, D., Schuster, D., Stumpf, F., Sigl, G.: Semi-invasive EM attack on FPGA RO PUFs and countermeasures. In: Proceedings of the Workshop on Embedded Systems Security, p. 2. ACM (2011)
55. Merli, D., Schuster, D., Stumpf, F., Sigl, G.: Side-channel analysis of PUFs and fuzzy extractors. In: Trust and Trustworthy Computing, pp. 33–47. Springer (2011)
56. Metz, C.: Microsoft bets its future on a reprogrammable computer chip. https://www.wired.com/2016/09/microsoft-bets-future-chip-reprogram-fly/ (2016). Accessed 02 May 2017
57. Microsemi: White Paper: Overview of Data Security Using Microsemi FPGAs and SoC FPGAs. Microsemi Corporation, Aliso Viejo, CA (2013)
58. Microsemi: EnforcIT Security Monitor. Microsemi Corporation, Aliso Viejo, CA (2015)
59. Microsemi: PO0137: PolarFire FPGA Product Overview. Microsemi Corporation, Aliso Viejo, CA (2017)

60. Moradi, A., Barenghi, A., Kasper, T., Paar, C.: On the vulnerability of FPGA bitstream encryption against power analysis attacks: extracting keys from Xilinx Virtex-II FPGAs. In: Proceedings of the 18th ACM Conference on Computer and Communications Security, pp. 111–124. ACM (2011)

61. Moradi, A., Oswald, D., Paar, C., Swierczynski, P.: Side-Channel attacks on the bitstream encryption mechanism of altera stratix ii: facilitating black-box analysis using software reverse-engineering. In: Proceedings of the ACM/SIGDA International Symposium on Field Programmable Gate Arrays, pp. 91–100. ACM (2013)

62. Moradi, A., Schneider, T.: Improved side-channel analysis attacks on xilinx bitstream encryption of 5, 6, and 7 series. In: Constructive Side-Channel Analysis and Secure Design—COSADE 2016. Springer (2016)

63. Morozov, S., Maiti, A., Schaumont, P.: An analysis of delay based PUF implementations on FPGA. In: International Symposium on Applied Reconfigurable Computing, pp. 382–387. Springer, Berlin (2010)

64. Nedospasov, D., Seifert, J.P., Helfmeier, C., Boit, C.: Invasive PUF analysis. In: Workshop on Fault Diagnosis and Tolerance in Cryptography (FDTC), pp. 30–38. IEEE (2013)

65. Oren, Y., Sadeghi, A.R., Wachsmann, C.: On the effectiveness of the remanence decay side-channel to clone memory-based PUFs. In: Cryptographic Hardware and Embedded Systems–CHES 2013, pp. 107–125. Springer (2013)

66. Pappu, R., Recht, B., Taylor, J., Gershenfeld, N.: Physical one-way functions. Science **297**(5589), 2026–2030 (2002)

67. Peterson, E.: White Paper WP468: Leveraging Asymmetric Authentication to Enhance Security-Critical Applications Using Zynq-7000 All Programmable SoCs. Xilinx Inc., San Jose, CA (2015)

68. Ravikanth, P.S.: Physical one-way functions. Ph.D. thesis, Massachusetts Institute of Technology (2001)

69. Roscian, C., Sarafianos, A., Dutertre, J.M., Tria, A.: Fault model analysis of laser-induced faults in SRAM memory cells. In: Workshop on Fault Diagnosis and Tolerance in Cryptography (FDTC), pp. 89–98. IEEE (2013)

70. Rührmair, U., Sehnke, F., Sölter, J., Dror, G., Devadas, S., Schmidhuber, J.: Modeling attacks on physical unclonable functions. In: Proceedings of the 17th ACM Conference on Computer and Communications Security, pp. 237–249. ACM (2010)

71. Rührmair, U., Sehnke, F., Sölter, J., Dror, G., Devadas, S., Schmidhuber, J.: Modeling attacks on physical unclonable functions. In: Proceedings of the 17th ACM Conference on Computer and Communications Security, pp. 237–249 (2010)

72. Rührmair, U., Xu, X., Sölter, J., Mahmoud, A., Majzoobi, M., Koushanfar, F., Burleson, W.: Efficient power and timing side channels for physical unclonable functions. In: Cryptographic Hardware and Embedded Systems–CHES 2014, pp. 476–492. Springer (2014)

73. Sahoo, D.P., Patranabis, S., Mukhopadhyay, D., Chakraborty, R.S.: Fault tolerant implementations of delay-based physically unclonable functions on FPGA. In: Workshop on Fault Diagnosis and Tolerance in Cryptography (FDTC), pp. 87–101. IEEE (2016)

74. Schlangen, R., Leihkauf, R., Kerst, U., Lundquist, T., Egger, P., Boit, C.: Physical analysis, trimming and editing of nanoscale IC function with backside FIB processing. Microelectron. Reliab. **49**(9), 1158–1164 (2009)

75. Schlösser, A., Dietz, E., Frohmann, S., Orlic, S.: Highly resolved spatial and temporal photoemission analysis of integrated circuits. Meas. Sci. Technol. **24**(3), 035102 (2013)

76. Selmke, B., Brummer, S., Heyszl, J., Sigl, G.: Precise laser fault injections into FPGA BRAMs in 90nm and 45nm feature size. In: 14th Smart Card Research and Advanced Application Conference—CARDIS (2015)

77. Simpson, E., Schaumont, P.: Offline Hardware/Software Authentication for Reconfigurable Platforms. In: CHES, vol. 4249, pp. 311–323. Springer (2006)

78. Stellari, F., Song, P., Villalobos, M., Sylvestri, J.: Revealing SRAM memory content using spontaneous photon emission. In: IEEE 34th VLSI Test Symposium (VTS), pp. 1–6. IEEE (2016)

79. Suh, G.E., Devadas, S.: Physical unclonable functions for device authentication and secret key generation. In: Proceedings of the 44th Annual Design Automation Conference, pp. 9–14. ACM (2007)
80. Swierczynski, P., Becker, G.T., Moradi, A., Paar, C.: Bitstream fault injections (BiFI)-automated fault attacks against SRAM-based FPGAs. IEEE Trans, Comput (2017)
81. Tajik, S., Dietz, E., Frohmann, S., Dittrich, H., Nedospasov, D., Helfmeier, C., Seifert, J.P., Boit, C., Hübers, H.W.: Photonic side-channel analysis of arbiter PUFs. J. Cryptol. **30**(2), 550–571 (2017)
82. Tajik, S., Dietz, E., Frohmann, S., Seifert, J.P., Nedospasov, D., Helfmeier, C., Boit, C., Dittrich, H.: Physical characterization of arbiter PUFs. In: Cryptographic Hardware and Embedded Systems–CHES 2014, pp. 493–509. Springer (2014)
83. Tajik, S., Fietkau, J., Lohrke, H., Seifert, J.P., Boit, C.: PUFMon: Security monitoring of FPGAs using physically unclonable functions. In: IEEE 23rd International On-Line Testing Symposium (IOLTS). IEEE (2017)
84. Tajik, S., Lohrke, H., Ganji, F., Seifert, J.P., Boit, C.: Laser fault attack on physically unclonable functions. In: Workshop on Fault Diagnosis and Tolerance in Cryptography (FDTC), pp. 85–96. IEEE (2015)
85. Tajik, S., Nedospasov, D., Helfmeier, C., Seifert, J.P., Boit, C.: Emission analysis of hardware implementations. In: 17th Euromicro Conference on Digital System Design (DSD), pp. 528–534. IEEE (2014)
86. Tehranipoor, M.M., Guin, U., Bhunia, S.: Invasion of the Hardware Snatchers: Cloned Electronics Pollute the Market. IEEE Spectrum (2017)
87. Trimberger, S.M.: Copy protection without non-volatile memory. US Patent No. 8,416,950 (2013)
88. Trimberger, S.M., Moore, J.J.: FPGA security: motivations, features, and applications. Proc. IEEE **102**(8), 1248–1265 (2014)
89. Tuyls, P., Schrijen, G.J., Škorić, B., Van Geloven, J., Verhaegh, N., Wolters, R.: Read-proof hardware from protective coatings. In: Cryptographic Hardware and Embedded Systems–CHES 2006, pp. 369–383. Springer (2006)
90. Verayo: Verayo PUF IP on Xilinx Zynq UltraScale+ MPSoC Devices Addresses Security Demands (2016)
91. Villasenor, J., Tehranipoor, M.: The Hidden dangers of chop-shop electronics: clever counterfeiters sell old components as new threatening both military and commercial systems. IEEE Spectrum (cover story) (2013)
92. Xilinx: Security Monitor IP: Industry-Leading Programmable Device Security Protecting IP and Mission Critical Data. Xilinx, Inc. San Jose, CA (2015)
93. Xu, X., Rührmair, U., Holcomb, D.E., Burleson, W.: Security evaluation and enhancement of bistable ring PUFs, pp. 3–16 (2015)
94. Yu, M.D.M., Devadas, S.: Recombination of physical unclonable functions. United States. Dept. of Defense (2010)
95. Yu, M.D.M., Hiller, M., Delvaux, J., Sowell, R., Devadas, S., Verbauwhede, I.: A lockdown technique to prevent machine learning on PUFs for lightweight authentication. IEEE Trans. Multi-Scale Comput. Syst. **2**(3), 146–159 (2016)
96. Zhang, J., Lin, Y., Lyu, Y., Qu, G.: A PUF-FSM binding scheme for FPGA IP protection and pay-per-device licensing. IEEE Trans. Inf. Forensics Secur. **10**(6), 1137–1150 (2015)
97. Zhang, X., Tehranipoor, M.: RON: An on-chip ring oscillator network for hardware trojan detection. In: 2011 Design, Automation & Test in Europe, pp. 1–6. IEEE (2011)

Printed in the United States
By Bookmasters